The Habitable Zone:
The Intersection of Science and Faith

by
William Richardson

ACKNOWLEDGMENTS

I thank my wife for her patience with me when there are so many different things happening simultaneously. I would like to thank my children, who are some of the best critical thinkers I know. I owe much to my father, who believes this book is vital at this time and who spent hours reading the manuscript and lending suggestions. I thank many students for questioning me, challenging me, and helping me pursue more accurate answers and perspectives. Though Dr. Hugh Ross does not know me, and we may differ slightly in our ideology, I would like to acknowledge his role in my journey. I attended a presentation of his in the early 1990's. His scientific perspective as an astrophysicist and man-of-faith had a tremendous impact on my life and my ability to help others. His research and writings are done with integrity; and his organization at Reasons to Believe is one of the most credible sources I have discovered for referencing what men and women in the scientific community are discovering about our life, our planet, and our universe.

ABOUT THE AUTHOR

The author is a statistician and instructor of 30 years. He possesses a B.S. in Mathematics and a M.S. in Applied Statistics. By 1984, his life had been profoundly and negatively impacted by misguided Christian teaching and decisions that arose from it. He walked away from church-life as a result. Many things bothered him, including the treatment of certain people groups and how to reconcile the extent of pain and suffering in this life if a loving God is truly in control. Like many, he simply could not accept a spiritual perspective that does not line up with reality. Tragedy in lives of others eventually led to him questioning and seeking again. In 1990 he began to investigate whether one could take a logical ascent to establishing the existence of a God, the existence of an afterlife, and the reason for so much pain and suffering. He soon recognized that the bias he and others have acquired over a lifetime makes it difficult to question, correct, and substantiate critical spiritual components. He experienced the cognitive dissonance that accompanies such a journey. For decades he has been involved in various efforts to improve the lives of others educationally, financially, emotionally and spiritually. But with his ability to connect spiritual dots while being accountable to universally accepted science and history, it is introducing others to the God of the universe that most compels him.

INTRODUCTION

Pursuing a spiritual ideology is an ongoing process. One can seek accuracy for a lifetime. It is challenging and time-consuming to develop an ideology in-line with science, history, and scripture. Everything should be questioned. Answers not available in one decade may become available in another. Human nature with its ego, bias, and self-centeredness [EBS] must be recognized for the distracting, deceiving force it can be. The good news is that a developing spiritual ideology can easily be adjusted when pertinent, accurate information is discovered. In this manner, one does not get lost in the dogma that can sidetrack or counterfeit ideas.

This book is the product of decades of seeking with such a process. It begins considering human nature and its potential to constrain one's ability to arrive at accurate perspectives. It makes a necessary, objective assessment of some popular scientific thought. Many who consider themselves scientific struggle to separate themselves from strongly held views of the 1970s. Though many fascinating new discoveries have been made in the decades since, there is a large group – including some of our popular science icons – that refuses to evolve with discoveries made. There are some who view science suspiciously, while other well-intentioned individuals construe science in a way that fits their ideology. Some so-called Christian scientists promote the impossible in the name of God. But anyone interested in integrity or in developing an accurate spiritual ideology should welcome and value scientific discovery. If a God truly exists, science should support his existence and not deny it. Otherwise, that God is deceptive or simply not very good at what he does.

Additionally, the book looks at 10 errant beliefs held by some Christians. Perhaps half of today's Christians possess one or more of these consequential and controversial beliefs. The negative effects are widespread and prevent many with integrity from considering certain spiritual ideas.

Finally, this book presents an ideology in line with science and in line with scripture. For various reasons, *scripture* is defined as the Jewish Tanakh - a holy book that is generally accepted by Christians, Jews, and Muslims – over half the world's population. The Jewish Tanakh are the scriptures as defined by the Apostle Paul and are the Law, the Prophets, and the Writings alluded to by Jesus. The Tanakh includes many scientific, historical, and prophetic references that suggest influence from a source outside our space-time dimension. While both science and spirituality have unanswerable questions and room for doubt, the ideology presented answers some of the most pressing questions people ask – including why pain and suffering are such necessary components of this current life.

DEDICATION

*Dedicated to the millions we have lost or harmed
due to our ignorance*

Table of Contents

The Trouble with Human Nature 1

Science: What We Do and Do Not Know 6

10 Errant Beliefs Held by Many Christians 17

The Fab Five: Abraham, Moses, David, Paul, and Melchizedek 44

Time and the Holy Book 52

The Law: Misunderstood but Vital 59

The Not-so-Fab Five: Pain, Suffering, Evil, Guilt, and Fear 71

200,000 Years of Humans 84

Signs and Prophecies 91

Ideological Suggestions 101

Chapter 1

The Trouble with Human Nature

The influence of human nature on political and religious views is profound. Two of our best weapons against our deceptive human nature are accountability to accuracy and a willingness to question ourselves.

There are important things to know about the human brain and how it functions. Our prefrontal cortex is engaged when carefully analyzing. However, the limbic region is where we tend to operate. It relies heavily on intuition, our immediate perception of things, and the meaning we have previously constructed to help make sense of what we are encountering. Daniel Kahneman won the Nobel Prize in Economics for his work on Economic decision-making. He displayed how expert financial analysts make irrational investing decisions despite being convinced the decisions are logical and wise. In his best-selling book Thinking, Fast and Slow, Kahneman compiled well-supported research to paint an accurate picture of how humans think and act. He refers to System 1, the limbic processor, and System 2, the prefrontal-cortex processor. System 1 is predominant. Unless there are certain external stimuli to engage System 2, System 1 gives permission and influence for System 2 to operate.[1] System 1 is very self-confident. Much like financial experts, this human flaw of being self-assured impedes everyone regardless of intelligence and level of education. All humans are susceptible to reaching inefficient and inaccurate conclusions. Awareness of this trait can help us better assess our ideology and suggestions of other ideas and theories. Unfortunately, high school curriculums rarely examine these critical characteristics of our being. Thus, most adults have little awareness of such things.

In addition, there are major motivators of human nature such as *yearning to belong* and the *avoidance of pain*. Belonging to a healthy relationship is powerful. However, most relationships are both good and bad, and people can rationalize compromising behavior or ideology for the reward of increased self-esteem. Belonging to a larger group of like-minded people can help us feel safe, but it can embolden shared, errant beliefs. Corporations, news sources, and other groups or individuals can exploit this need-to-belong and thus influence beliefs. Most would agree that where we get our news and who we interact with can have a major impact on our ideology. Many naturally assume it's the *other guy* that is being duped. Our tendency to *think it's the other guy being duped* is perhaps the best example of our human nature at its core. When we find others who agree with us, we can easily overlook whether our position is supported by evidence and reality. People tend to be very self-assured, and so much of our information-intake produces confirmation bias. The political landscape is viciously divided. Without a general understanding of human nature and how to defend against it, it may be difficult to successfully emerge from what today must be considered a political crisis.

How human brains came to be hard-wired this way is not certain. Naturalists would suggest an evolutionary explanation. Over millions of years, species had to operate intuitively and instinctively to protect against immediate danger from predators and nature itself. This would explain why this quick-thinking, intuitive processing is primary in us. We quickly assign meaning to the things we see. The meaning comes from our past experiences and our own bias, and we are very convinced of our decision-making. Alternatively, many with a religious ideology would claim that human nature is the *sin-nature* God created in us. The idea might be that God created humans with this flawed nature, placed them in a beautiful, yet brutal world, and provided opportunity to discover the primary purpose of this life for people willing to employ humility and integrity to overcome human nature. Perhaps our condition is an orchestrated result of both explanations and more.

Regardless of how we arrived at our current state of existence, there is much agreement within the scientific community of how the two regions of our brain interact and process. In the future, perhaps the education system will better inform our children on the realities of our human nature and how we might overcome it to achieve and to understand at the highest levels possible. In many classrooms of the past and present, a teacher is in front of a captive audience telling them what to think – providing the *correct* information. From a young age, people are programmed to trust the experts in front of them. In History class, for example, instructors and documentaries can look back in time with seemingly perfect vision and share just the right factors and events that led a society into war or financial crisis. Students may emerge with notions of the signs to look for in any situation that will point to economic demise or war. But life is far more complex with wide arrays of factors continually interacting. Looking back on history provides much more clarity than making history. If recognizing warning signs was so easy in real time, our history books would not be filled with so much destruction and turmoil. And **always**, the greatest contributor to history is human nature.

There are many examples of negative *group-think* via human nature.

The Habitable Zone

In the 1960's and 1970's, many argued that it was safer to not wear a seatbelt in a car. There was anecdotal evidence of a person thrown from a car and surviving when he would have died if wearing a seat belt. It took years and government mandates for sound, scientific and statistical wisdom to win the debate. Social media and the internet tend to make human error, confirmation bias, and thus *group-think* more prolific today than ever. Many love conspiracy theories. In the 1960's a writer for Playboy Magazine teamed with an author of a parody religion to create fictitious letters from readers for and against the existence of a secret society called the Illuminati.[2] They knew the actual Illuminati began in 1776 and died out soon after. But their hoax gained traction, and today millions of people believe this fictitious secret society lurks behind the scenes. Currently, the arguments for and against climate change and the role of mankind in it are riddled with the influence and consequence of human nature.

A powerful perspective on life that can help us avoid pitfalls is the practice of seeing everyone and most ideas as both *good **and** bad* as opposed to intuitively *good **or** bad*. People like to categorize others as heroes or villains. This is dangerous. Even those who are primarily good will make bad decisions. Without recognizing this, we leave ourselves vulnerable to being harmed and overly-affected by those we trust. Recognizing this makes it far easier to implement better understanding and change. Individually, it is powerful to view one's self as both *good **and** bad*. It removes much of the unnecessary pressure to always do good and be good. All of us have traits we despise. And all of us have some inaccurate perceptions of things. When we can identify these, it is not an indication that we are wrong, or that we are failures, or that we are bad. It simply implies opportunity to add more good to our goodness and delete some bad from our badness. Awareness of human tendency and the willingness to question one's self are valuable assets in the quest for wisdom.

Chapter 2

Science:

What We Do and Do Not Know

"Great spirits have always encountered violent opposition from mediocre minds. The mediocre mind is incapable of understanding the man who refuses to bow blindly to conventional prejudices and chooses instead to express his opinions courageously and honestly."
Albert Einstein

Here is a statement that is contrary to what most learned in school or witness in some documentaries. It may seem like it shouldn't matter, when it is quite critical. The statement is this: based upon the sum of current research from investigating astrophysicists, chemists and biologists, we would have to conclude that animal life is not at all abundant in our universe. While there could be a fair number of planets capable of supporting life if animal life were transported there with the proper additional resources, there is likely not even one other planet in the known universe that would have the hundreds of necessary characteristics similar to Earth's that would provide it the protections and resources to accommodate 4 billion years of preparation and evolution to produce advanced animal life such as we have here unless something from outside of our space-time dimension intervened. With limited information in the 1970's, it was plausible to suppose that millions or even billions of other advanced civilizations exist in our universe due to random chance. But that was decades ago. We know today that a planet requires so much more than some water, some land and a steady burning sun. While millions of intelligent people follow the religious *group-think* of a 6000-year-old Earth, tens of millions of intellectuals hold to the academic *group-think* that there are millions or billions of other civilizations on planets elsewhere in the universe. We were taught such a thing growing up, and this notion of life elsewhere has not evolved as science has provided more clarity.

Ultimately, for a hypothesis to be taken seriously, it must be supported with probability theory. Some people dismiss probabilities out of sheer ignorance. Others know just enough about probability to be dangerous and continue to promote ideas with great confidence, when in fact a more accurate mathematical and scientific approach would help them to recognize the frailty of their arguments. This happens more frequently than one might realize among people possessing advanced degrees. In fact, it is *conditional probability* that tends to hold the key in supporting a hypothesis or ideology.

Some of the more intelligent misunderstandings of probability include the following. "We are here, so there must be an explanation for why we are here." And then the individual assumes that the explanation includes only sources from within our space-time dimension. Such is the foundation of Naturalism, a philosophy that is not pure science but is what we have taught as science in American public schools for decades. Another intelligent misunderstanding of probability proceeds something like this: "We know our very existence on Earth defies the odds, but with millions or billions of other planets like Earth in the universe, it is logical to understand that at least one of those chances would produce a planet like Earth with its incredible life forms." With this second erroneous conclusion, many highly intelligent and mathematically inclined individuals excuse probabilities and resort back to the first intelligent misunderstanding – thus assuring themselves that there is no God, and that it is very logical to conclude the presence and progression of life on Earth occurred via random chance and without the need of an overseer. Of great concern in academia is the assuredness with which many assume we will find natural explanations; and worse yet, a lack of awareness of contradictions or gaps along the chronological, natural explanations of the past 4.5 billion years on Earth.

Here is a summary of what has taken place on Earth in the past 4.5 billion years:
Complex life on our planet required self-replicating material that could produce life and then allow the replication of life. While many assume that this self-replicating material could emerge four billion years ago from substances contained on Earth, the evidence points to a different conclusion. Protocells have been touted by some as a potential intermediary step between prebiotics and origin of life; but thus far in the scientific search to form protocells, self-replicating material is required.[1] So while protocells may be an intermediate step, they do not explain the origin of self-replicating material. There are many accomplished biologists who do not grasp this point. Some respond with statements like, "It is easy to copy and reproduce DNA." But they do not realize it is easy to do such a

thing *today*, because *today* we have DNA. With it already in existence, we can easily copy it. We are speaking of the conundrum regarding how Earth first produced DNA or any other self-replicating material four billion years ago. The one who can answer that question and support it may well receive a Nobel Prize. This is because we do not yet have an answer. Many researchers are hard at work trying to determine the origin of nucleotide bases and of self-replicating material. In 2016 NASA scientists reported that simulated conditions of ice samples on meteors travelling in space at -440 degrees Fahrenheit can, when subjected to UV rays, break down and form into uracil, thymine and cytosine.[2] The problem is that, while guanine, adenine and uracil have been found in meteorites, cytosine and thymine have not.[3] Pearce and Pudritz [2015] found that cytosine fails to survive within meteorite parent bodies as a result of aqueous deamination; and thymine does not survive when H_2O_2 is present because of an oxidation reaction between the two.[4] As fortunate as our planet is, it did not have the capacity as a planet to produce life without the building blocks of life being transported here undamaged from elsewhere. And once transported, somehow the bases would bond and come together. That is why scientists are simulating conditions in nearby space in hopes of determining how this very fortunate emergence of prebiotics and life could take place on Earth. The more scientists uncover and determine, the more some scratch their heads and recognize how extremely lucky Earth is to possess advanced animal life.

There is a plethora of elements on our planet, but initially a vast amount of these were water-soluble. This means that water on Earth nearly four billion years ago was poisoned with high levels of lead, zinc, copper and other elements. Prokaryotic bacteria needed more than a billion years to convert these water-soluble elements to non-water-soluble and to generate an ever increasing volume of atmospheric oxygen.[5] A huge and beneficial consequence of the bacteria accomplishing this is that vast deposits of iron ore and copper and gold would be available billions of years later when homo sapiens arrived, so that we could eventually create the technology we enjoy today. A billion years ago more complex life forms emerged in addition to bacteria, including sponges and worms (and other larger species that were wiped out by a mass extinction prior to the Cambrian explosion). Four more nicely-timed and beneficial mass extinctions, each followed by swift repopulations and ecosystems, have occurred since the Cambrian.[6] Those events helped to provide a large supply of fossil fuels waiting for humans at just the right time. Additionally, given the sun's changing luminosity, had it not been for these mass extinctions and repopulations at ideal times, there existed the possibility that Earth might otherwise have been permanently sterilized.[7] Perhaps the five major extinction events are rarely discussed in schools due to the problems they present in attempting to promote a smooth evolutionary model embraced for decades now. The emergence of the new ecosystems – these explosions of life – behave as if anticipating the soon-to occur atmospheric changes – something that makes more sense if a god orchestrated the change. Natural change tends to be reactionary rather than anticipatory, which is why these more *anticipatory* extinctions and repopulations lend credibility to influence from a source beyond our space-time dimension as opposed to strictly terrific luck and timing if only nature is responsible.

For a planet like Earth to have the billions of years to spawn, develop, and nurture life, it needed protection from harmful rays that would otherwise sterilize a planet and cause it to become barren. This requires a sustained electromagnetic field. In order to accomplish this, radioactive elements are necessary. The decay of radioisotopes within our planet maintains a high enough level of heat in our core to keep iron and nickel molten and spinning in order to uphold the electromagnetic field. Radioactive isotopes also generate tectonic activity required to sustain life. Water is not rare in the universe. But Earth's proportional composition of elements, minerals and water are quite rare and quite necessary for our planet to support and promote life.[8] It is not until the nine billion-year mark of the universe that the abundance of radioactive isotopes and other heavy elements are sufficiently present in the universe along with key elements like flourine.[9] Thus, Earth formed at a nearly ideal time in the universe for acquiring the heavier elements it possesses. It is not that Earth-like planets were impossible to form before the nine billion-year mark. It is that it was far less likely. And arguably, without all these elements finally present at the nine billion-year mark, animal life might be impossible or nearly impossible.[10]

For our sun and its forming planets to grab these elements, our solar system must have formed in a more dense location of the galaxy, where enough explosions of various stars provided elements like fluorine in addition to radioisotopes and the other heavy elements.[11] However, we could not stay in that same location with so many forces at work. We needed to end up between the arms of the galaxy – in our galaxy's habitable zone.[12] Hence, it is not just that the plethora of heavy elements were finally available in the universe. It is that a planet and its system formed in a location where it could grab all these elements and then move to an area without so many rays and forces at work that would otherwise prevent animal life from developing. And Earth again defies the odds by not being so near the sun with its high mass density and composition of heavy metals. We actually have a higher mass density than Mercury, yet are located further away from the sun – in the animal habitable zone of our planetary system.[13] Earth seems to catch every lucky break.

Earth needed correct timings of increases and decreases in certain minerals and gases. The Great Oxygenation Event 2.4 billion years ago was a result of a nickel famine that led to a significant decrease in methane, that in turn allowed for an increase in atmospheric oxygen.[14] Yet, if this chronology of events had occurred earlier, when the sun's luminosity was lower, Earth could have transitioned into a long-term ice ball.[15] Mathematically, the timings of so many key interactions over four billion years continues to signal the rarity of a planet able to coordinate the distribution of factors in order to produce the necessary progression toward more sophisticated life. It is as if a God knew what conditions – like the sun's luminosity – were right around the corner and intervened with mineral and gas interactions, mass extinction events followed quickly by new ecosystems, and the coordination of other factors all in preparation for humans millions and billions of years later. Animal life is not at all common in this universe. Too many things had to occur and interact in a just-right fashion.
In 1961, Dr. Frank Drake introduced his Drake equations, with

which he attempted to estimate the number of advanced civilizations in our galaxy.[16] Based upon the scientific evidence of his day, there was no reason to doubt his conclusions that billions of civilizations should exist in our galaxy alone. In the 1970's and 1980's, Carl Sagan also proposed that there must be many civilizations within the universe with "billions and billions" of stars out there. Sagan was more than an excellent scientist. He was an outstanding human being. It is tempting for us to hold to his beliefs because of his appeal. However, subsequent scientific discovery no longer supports his contention.

In 2016, NASA reported that it found an additional 1284 exoplanets. But only nine were adequately rocky and the right size and in the habitable zone.[17] That is less than 1%. And there is a difference between a planet that is capable of "supporting life," if existing life was transported there, and a planet capable of spawning, nurturing and producing advanced, technological life such as what Earth accomplished. Mike Wallace, Space.com senior writer, quoted NASA scientists as saying the following, "About 550 of the newly validated 1284 [exoplanets] are small enough that they could be rocky, scientists said. And nine of those might be able to support life."[18] Balanced information like this should be shared in colleges and high school. But such information rarely finds its way into the classroom. True scientific pursuit helps us recognize the need for much more than merely a small rocky planet, with some surface water, abiding within a habitable zone of a solar system and galaxy.

Consider Earth's need for an exceptionally large moon. Our moon is literally one-in-a-trillion. There is still some uncertainty as to how Earth captured such a large, ideal moon with its gravitational force. The most likely natural explanation might be a collision with our moon as both were forming.[19] We have an ideal 23.5-degree tilt-of-axis to our sun and an ideal spin rate today as a result. This tilt gives us a much milder climate in a far greater swath of our planet. The fact that our moon is so massive relative to Earth means that it is able to create strong enough tidal activity to cleanse ocean waters from toxins and replenish them with nutrients; and it has the gravitational force to hold us at our ideal tilt and not allow the sun's gravity to yank us more upright and slow our spin rate down – which would lead to harmful climate changes.[20] Our spin rate is very favorable to life. Spin too slow and experience too great of a temperature variance daily, leading to a dramatic impact on plant and thus animal life.[21] We could not have expected to capture such an ideally large moon in our gravitational pull had it not been for this precise collision.[22] Finding other planets with sufficiently large moons relative to the size of the planet has an extremely low probability. It is not enough to be one of the relatively few small, rocky planets with a thin crust, an ideal proportion of water and located in a habitable zone. The planet needs a large moon like ours to be a candidate for lower life forms evolving to higher life forms.

As our age of discovery emerged, we recognized another coincidental benefit from this moon. It just happens that our moon is 1/400th the diameter of the sun and 1/400th the distance from Earth to the sun. This allows for occasional total eclipses – the moon nearly perfectly blocks the sun and allows us to analyze. The observations made by scientists during total eclipses of the past 200 years have revealed the chemical make-up of bodies in space, confirmed Einstein's theory of general relativity, and allowed us to learn much regarding the universe while standing firmly on our planet. The incredibly coincidental ratio of the diameter of the moon to the diameter of the sun and the ratio of the distance of the moon and sun from Earth has allowed this to happen.

One of the easiest requirements to comprehend regarding protection of a developing planet for billions of years is the need to have four giant planets in the solar system whose gravitational pull can divert asteroids and comets away from colliding with Earth. No matter where we are in our orbit of the sun, Jupiter, Saturn, Uranus and/or Neptune are protecting us by diverting huge space debris. Without at least four giant planets in the same solar system, a planet could not accomplish what Earth has accomplished. The asteroid strikes or collisions with comets would be far too frequent and too severe to allow the billions of years necessary for life to progress. In 1994 fragments of a comet were pulled into Jupiter with significant force.[23] There is always at least one of these giant planets that has our back. Additionally, our tiny blue planet is fortunate that the asteroid belt is contained where it is.

In addition to requirements most of us would understand, there are many more requirements being discovered by astrophysicists and chemists in specialized areas. These include things like "the infall velocity of matter into the dark matter halo of the potential life support galaxy," and "the average rate of migration of aqueous fluids through the planet's upper crust." These and more than 900 other documented requirements can be found in the appendix of the book, More Than a Theory by Hugh Ross [2009]. As persons ponder whether animal life on Earth is a result of purely natural factors or if there has been influence from a source beyond our space-time dimension, it is important not to reach unsupported conclusions. Just because science does not currently have explanations for things like the origin of self-replicating material does not imply that it must be a God who designed it. While it is tempting to reach such conclusions, one must accept the uncertainty and the role of faith in various answers. Conditional probability is critical in exploring what causes are more plausible than others. The assertion I make and that is considered later in this book is that there is evidence we have been contacted by an entity not bound by our dimension of time. Because of this and because of the rarity of a planet with the more than 900 characteristics necessary for advanced animal life, it is compelling to put one's faith in the influence of this entity.

Returning to the statement that "We are here, so there must be an explanation," and to the statement, "We know that our very existence on Earth defies the odds, but with millions or billions of other planets like Earth in the universe, it is logical to understand that one of those chances would produce a planet like Earth with its incredible life forms," we now should recognize the second statement is not supported as more and more requirements for life are discovered. The first statement is true, but it is not logical to *assume* the reasons are strictly from forces within our space-time dimension. Uncertainty exists, and faith plays a role in the list of causes we accept.

Chapter 3

10 Errant Beliefs Held by Many Christians

Paul warns religious individuals who rely too much on the Law, who are guilty of judging others, and who think too highly of themselves: "God's name is blasphemed among the Gentiles because of you." Romans 2: 24

Various documentaries have been produced investigating issues surrounding the life and death of Jesus. A valid point is often made regarding the lack of evidence for many events claimed in the gospels. While it is true that evidence is not overwhelming in some cases and that some details cannot be substantiated, it is also true that evidence is more than sufficient to establish who Jesus was and the many signs and prophecies that he fulfilled. One historical figure – perhaps more than any other – provides the greatest insight into who Jesus was and how he accomplished what was planned for him. Saul of Tarsus was a Roman citizen and a highly educated Jewish Pharisee. [New International Version, Phil 3: 4 – 6] Saul despised Jesus and his followers – considering them to be heretics. He rounded up early Christians to be imprisoned or put to death. But then something remarkable happened. Through various events and ultimately through his knowledge of Jewish scripture, Saul clearly recognized Jesus as the Jewish Messiah. Saul converted, changed his name to Paul, and became the greatest instructor of the Christian faith. Early Christians feared this man who had destroyed lives of many early followers. But over time it became obvious that his conversion was genuine and profound. The Christian New Testament contains many of Paul's instructional letters. He gave up his wealth and stature in order to travel much of the known world and explain why Jesus was the promised Messiah. He was dangerous in the eyes of powerful Jewish leaders, who rejected Jesus as the Messiah. Thus, Paul was a wanted man – escaping death or capture on various occasions. After many years of teaching, church-planting, and missionary work, he was captured and ultimately beheaded for his faith. Of additional significance is the fact that none of the twelve disciples of Jesus recanted. All but one chose to be put to death by his captor rather than retract the message regarding Jesus. These men were convinced of the resurrection.

Paul's instruction is extremely relevant today. There are foundational ideas that many churches ignore. One example is the common interpretation of 2nd Timothy 3: 16. It is part of a letter from Paul to a young Timothy, who was a significant figure in the early church. The verse states that, *"All scripture is God breathed."* Many Christians claim this pertains to the entire Bible – Old and New Testament. However, Paul clearly identifies what is meant when he references all scripture. Here were his words two verses earlier.

"But as for you [young Timothy], continue in what you have learned and have become convinced of, because you know those from whom you learned it, and how from infancy you have known the holy Scriptures, which are able to make you wise for salvation through faith in Christ Jesus. All scripture is God

breathed and is useful for teaching, rebuking, correcting and training in righteousness, so that the man of God may be thoroughly equipped for every good work." (2Tim 3: 14 – 17) Here Paul defined what the scriptures were. They do not include the New Testament of the Bible, since it had not been written when Timothy was growing up. The scripture Paul references is the Jewish Tanakh (or Tanach). It is the Jewish acronym for the Torah (Pentateuch, which includes the Law), the Navi (Prophets), and the Ketuvim (Writings, including the Psalms). Luke 24: 44 records Jesus referencing the Tanakh when he stated, *"Everything must be fulfilled that is written about me in the Law of Moses, the Prophets and the Psalms."* It is quite significant that the scriptures be correctly identified as the Tanakh, because all the Abrahamic faiths (Judaism, Islam, and Christianity) assign special meaning and inspiration to the Tanakh - though Islam does not consider the entire Tanakh to contain truth. Delving deeper into Paul's teachings, he claims that the death and resurrection of Jesus paves the way to eternity for all people groups - not just Jews and professing Christians. (Rom. 2. 12 – 16)

The New Testament must be looked upon differently than the Tanakh. In the four gospels, it contains various perspectives of the life of Jesus. The Book of Acts is a record of the early church. Paul's letter to the Romans is basically a dissertation on Christianity; and his other letters include what he claimed the scriptures provided: *"teaching, rebuking, correcting, and training in righteousness, so that every man of God may be thoroughly equipped for every good work."* Indeed, Paul's letters rebuked groups who were out of line, encouraged and instructed others, and had a common theme of producing wise and effective followers of Christ. Paul is uniquely qualified as a scholar and as one who challenged his own **EBS** [ego, bias and self-centeredness] to recognize who Jesus was and how the ancient Tanakh outlines God's plan for all mankind. Letters from other key figures and John's Book of Revelation are also part of the New Testament. Applying literary integrity is crucial in arriving at an appropriate ideology.

When one encounters a seeming contradiction in scripture or the New Testament, it is wise to pay closer attention to it and evaluate it as opposed to throwing the idea out. This is because some seeming contradictions hold the key to important discovery once they are carefully investigated. One seeming contradiction was recorded in Matthew's Gospel. Jesus had been beaten and mocked by soldiers. It was reported that they had encircled him and cast lots for his clothing. He was ultimately hung on a cross, where his hands and feet were pierced with nails. Matthew records Jesus calling out to God. *"My God, my God, why have you forsaken me?"* (Matt 27.46) This is a strange thing to write, as it appears to suggest that Jesus is a fraud. However, it leads those interested to Psalm 22 written by David 1000 years earlier and recorded in the Tanakh. The Psalm begins, *"My God, my God, why have you forsaken me?"* Psalm 22 goes on to describe events in the final moments of the life of the Messiah to come 1000 years later – not in the life of David, himself, as none of these things happened to David. It mentions that *"a band of evil men has encircled me, they have pierced my hands and feet. I can count all my bones; people stare and gloat over me. They divide my garments among them and cast lots for my clothing."* (Psalm 22.16 – 18) Paul, in his writings to the Galatians, then brings us full circle through prophecies and declarations in the Law to explain how God was able to curse Jesus with the wrongdoings of all mankind. This will be discussed in more detail later. There is no good reason why a spiritual ideology requires one to dummy down. On the contrary, when science, history, math and a holy book – influenced by a source beyond our space-time dimension – are in harmony, there is a peace that passes all understanding.

Human nature is complicated, and everyone should be seen for what they are – both good and bad. Christians, like any other group, are on a continuum from mostly good to fairly bad. There are incredible Christians who quickly earn the respect of those they encounter; and there are Christians who hold to and verbally share ideas with roots in current, popular Christendom as opposed to healthy, supported Christian doctrine. Millions of people fall into this second group but may be convinced that their views are accurate and may be convinced that they are victims on the world's stage. The role emotion plays in belief systems should not be overlooked. Inaccurate beliefs may be held with absolute conviction when the underlying assumption is "God's word says this," and when there is pressure from group-think of churches, on social media, and in every day communication. Even when some of these individuals are shown the scripture and how it deviates from their beliefs, the years of indoctrination and peer pressure and assumption literally convince them that their views are correct. Some of the most respected evangelists are among this group. It does not make them bad people. It makes them human. Just because someone is religious does not imply they have the exceptional level of humility to wrestle completely with their beliefs. The minute one assumes to be correct and avoids the pain of challenging beliefs, they follow the path of most humans over time.

Though many of us know people who are fantastic ambassadors for Christ, many inside and outside of religious circles are aware of some disturbing ideas promoted by some churches and some professing Christians. In addition to disturbing trends, there are subtle ideas that lead to undesirable consequences. Therefore, with accountability to the Tanakh, the Apostle Paul and history and science, the following is a list of 10 common errant beliefs that need to be addressed. It is no exaggeration to say that millions of lives depend on correcting these reality deviations.

10 Errant Beliefs That Need to be Addressed:

1 The belief the Earth is less than 10,000 years old

2 The belief that *days* mentioned in Genesis are each 24 hours long

3 The belief there was a global flood – not a local one – and that God instructed Noah to take two of every animal aboard an ark

4 The belief that God reacted to Adam's fall and only then devised a plan to come to Earth as a man and die for our sins

5 The belief there was no physical death before Adam sinned

6 The belief that Christians should aggressively confront and shun the LGBTQ community

7 The belief that most American Muslims are in favor of radical terrorism to bring down the West

8 The belief the devil is solely responsible for the presence of pain and suffering and evil

9 The belief God is too nice and too loving to create a hell

10 The belief those outside of Christianity cannot enter heaven

1 The belief Earth is less than 10,000 years old
  Multiple scientific factors point to a universe that is more than 13.7 billion years old. Scripture [the Tanakh] does not provide a complete time account. Two of the gospels in the Christian New Testament, Matthew (Matt. 1.1 – 16) and Luke (Luke. 3.23 – 38), list a genealogy from Abraham to Jesus to show that Jesus is a descendant of Abraham. But comparing the two genealogies exposes glaring differences. This appears to be a significant contradiction between two of the New Testament books of the Bible. Recall that exploring contradictions often holds keys to bigger ideas. Explaining this seeming contradiction helps uncover the key to a biblical age of the earth.
Believing these genealogies are comprehensive is a natural assumption on the part of a reader. In other words, the reader assumes that every ancestor between Abraham and Jesus is listed. However, a little math reveals these genealogies are nowhere near complete. Generations were less than 20 years. These genealogies cover a period of 2000 years. There would need to be over 100 names [generations] on these two lists to cover the entire 2000-year span. There are not. Since both lists skip a majority of ancestors, it is easy to understand why differences exist. Therefore, this is not a true contradiction

bringing into question the validity of the gospels. Rather it is a recognition that lists of Hebrew genealogies do not include every generation.

When Bible scholars claim the Earth to be 6000 years old, they fall into the trap of assuming the genealogies from Adam to Noah and from Noah to Abraham in Genesis are complete with no skipped generations. [See chapter 8 for more information on homo sapiens present on Earth for the past 200,000 years.] Just like the genealogies in the gospels were not complete, neither are these. Looking to the notes in most study Bibles for the meaning of the Hebrew word 'abba' shown for Genesis 11: 10, one finds that the word can mean *father or ancestor*.[1] The only way this genealogy can match history and science and thus be accurate is if, like the genealogies in the gospels, it is an incomplete list of ancestors; and generations were skipped for whatever reason. Since the alternative converts the story to fiction and would condemn the entire Bible and Tanakh for having glaring errors, we should pursue the notion of skipped generations and allow history and archaeology to line up with scripture and to estimate the time for Noah. More on this when we deal with Errant teaching number 3.

But the conclusion is this: The Tanakh, and thus The Bible, does not contradict a 13.7 billion-year universe.

2 The belief *days* mentioned in Genesis are each 24 hours long
Since many Christians with the best of intentions (a perceived adherence to scripture) teach of only 4000 years between Adam and Jesus, they tend to accept the days of creation in Genesis to be 24-hour days. However, the Hebrew word translated day is the word *yowm* [or *yom*]. It can also mean a long period of time like an age of time.[2] There is an obvious contextual clue in

scripture that *yom* is a long period of time and not 24-hour days. The Jewish day begins at sundown. In Genesis chapter one *"there was evening and there was morning"* is used when describing each of the first six *yom* [days] of creation. But not so for day 7 at the beginning of Chapter 2, where we are told that God rested from his creation [not rested for good]. There is no mention of evening and morning for day 7. This would imply that we are still in the seventh *yom* – period of time. And a very cool thing happens if we read Chapter 1 of Genesis with the possible interpretation that each day is a long period of time. The passage in Genesis could then be describing the big bang and development of our universe that has occurred over 13.7 billion years . [see Chapter 5 on Time and a Holy Book for a thorough discussion of Genesis describing the big bang and the correct progression of cloud cover breaking up to eventually reveal the sun and stars.] Scripture and science agree here. Teaching a young Earth – as many Christians do - is extremely detrimental. This implies that scientific discoveries - different from scientific theories - cannot be trusted. It is detrimental and potentially dangerous when a religious organization teaches that we should ignore/disrespect scientific discovery and then teaches doctrine that contradicts scripture and history.

3 The belief there was a global flood – not a local one – and that God instructed Noah to take two of every animal aboard an ark
The Tanakh [Bible] does not describe a global flood. It describes a localized flood – likely, the entire Mesopotamian Plain. And it is not true that God instructed Noah to take two of every animal. Chapter 7 of Genesis is where the account of the flood

begins. In verses 2 & 3 God told Noah to take 7 pairs (a male and female) of certain animals, 2 pairs of others, 7 pairs of birds, and by implication none of other animals. A look at the meanings of the Hebrew words used helps determine that only certain types of animals were needed aboard the ark. Livestock for when Noah's family started over, animals for sacrifice, and birds were among the necessary animals on board. This is because the flood was not global. In Psalm 104, there is a poetic description of creation and the progression of Earth. It declares that, after the land mass came up out of the water initially on Earth, God would never again allow the tops of the mountains to be covered with water. Thus, a global flood in Noah's day would contradict this scripture. Due to this and the contextual clues in the Genesis account, the flood was localized and did not cover the entire planet.

Combining what is and is not possible according to scripture with what archaeology and other branches of science reveal about humans the past 200,000 years can lead to a few possible scenarios regarding a flood affecting some or all humans and where and when it could have occurred. This is discussed in Chapter 8.

4 The belief that God reacted to Adam's fall and only then devised a plan to come to Earth as a man and die for our sins
Regarding the Tanakh and New Testament, there is a huge problem with many Christians who teach that this plan was only devised after Adam sinned. The biggest problem would be the suggestion that God can be fooled and is not in control.
Genesis 1:1 says *In the beginning*. This implies that our space-time dimension (our universe) had a beginning [out of nothing that can be seen]. This is scientifically accurate. Thus, God

exists outside of our space-time dimension. Genesis mentions that God declared *"Let us make man in our image"* (Gen. 1.26), establishing the multiple persons of the godhead - namely the Father, the Son and the Holy Spirit. The Apostle Paul tells us in 2Timothy 1:9 that the grace given us in Christ Jesus was given to us *before time began.* Thus, Paul implies that the plan for the Son to redeem us through his sacrifice was established before our universe began 13.7 billion years ago. Because God exists in such a realm outside our space-time dimension, He exists in our past, present and future simultaneously – *the same yesterday, today and forever.* This is how God is in control while giving us a free will. And thus, God is never fooled. He sees and is *in the future.* This notion that he initially constructed this universe to be eternal but then Adam sinned and messed up everything is not what Paul teaches; and it leads logically to a very dangerous conclusion. It suggests that God is not in control: He could be fooled or surprised. Accepting that God reacted to Adam's fall and did not anticipate it leads to many contradictions to reality, scripture, God, and the plan. Additionally, the laws of Physics suggest this universe is finite. Left to itself, it will eventually run out of energy and cease to exist. This universe was not intended to be eternal. Again, scripture and science agree. God works on an eternal habitat for us in another space-time dimension.

We have become a society of people that seem to be okay with fanciful spiritual perspectives that can go in any direction without a requirement of accountability. Hence, many assume they are correct. People despise others for their beliefs, argue over beliefs, refrain from discussing beliefs and mock spirituality. This is all very needless, illogical and detrimental; but it is often how we humans behave.

5 The belief there was no physical death before Adam sinned
Some Christians interpret references to spiritual death in Genesis as being physical death. They preach that nothing died before the Fall (of Adam) and combine this idea with the notion that the Earth is extremely young. This is not in line with accurate scriptural interpretation and ignores the wealth of evidence for various life forms. We have evidence of sponges and worms that existed over 560 million years ago. And the fossil fuels that have led us into this industrial age are the wonderful consequence of death on Earth for millions of years before homo sapiens emerged. For these and other reasons, many intellectuals feel there is no valid reason to consider the Christian faith when so many ideas deviate from what is known. I feel badly for some young Christians who are ridiculed in classrooms and universities for standing up for a *young-earth* ideology. They are displaying their commitment to God – a brave and loving action. Yet, more harm than good comes from this. Thankfully, God is not a deceiver. He did not create the universe to look a certain way and then inspire a holy book to contradict it all. If we heed the teachings of Paul, read scripture with literary integrity, and utilize scientific discovery to put things into context, we realize the harmony between the Tanakh and reality.

6 The belief that Christians should aggressively confront and shun the LGBTQ community
In many Christian circles, believers are instructed to immediately cut off communication with anyone found to be gay and not to accept them back until the person repents and changes. In some circles, those identified as LGBTQ are

encouraged or even forced into therapy to convert them to straight. These are actions that most point to the severe lack of understanding among many professing Christians. Self-righteous behavior is a part of our nature that we are to battle against. Instead, it is being encouraged. Jesus condemned religious leaders in his day for such behavior. The outward condemnation of the LGBTQ community by the religious right is evidence of a disconnect from what Jesus and Paul would propose as the foundation of our righteousness – namely, that it comes from faith (as in the case of Abraham) and grace (for those of us who admit our humanness). To ignore our own humanity and to miss the point being made by the Apostle Paul in the first two chapters of his Letter to the Romans, is what Paul warns of in the second Chapter of Romans where he states that *"God's name is blasphemed among the Gentiles because of you."* (Rom. 2.24) Many love Romans Chapter 1, where some seem to think it encourages the open condemnation of all in the LGBTQ community. But Paul continues his thought in Chapter 2, *"You, therefore, have no excuse, you who pass judgment on someone else, for at whatever point you judge the other, you are condemning yourself."* The teachings of Paul and Jesus are consistent. In Matthew 7:1 and 2, Jesus is quoted as saying, *"Judge not, or you will be judged. For in the way you judge others, you will be judged, and with the measure you use, it will be measured to you."*

Most of us can recognize perversion when we see it - the (child) pornography industry, human trafficking, incest. The description of the cities of Sodom and Gomorrah are sobering. Common place were heterosexual and homosexual orgies and the threat of assault outside one's door. Some think that it was the homosexual perversion that most typified the place; and

29

therefore, it is homosexuality that we most need to defend against. From a logic and math standpoint, that is not a valid conclusion. The explanatory variable [root cause] was runaway lust. The response variable [the result or action from runaway lust] was outward displays of immorality of both a heterosexual and homosexual nature. And most of the laws given to Moses involving lust and immorality pertain to heterosexuality – not homosexuality. Homosexuals did not corner the market on sexual sin.

For the sake of discussing life today, there needs to be a distinction between behaviors within the LGBTQ community. Most can agree that a person who progresses in perversion and needs to be with multiple partners and/or commit indecent acts is a depraved person. This is true in the LGBT community as well as within the heterosexual community. However, the majority in the LGBT community practice monogamy and have an emotional and sexual bond to one or no partner based upon respect and trust, including many young people who identify this way.

The primary text in the Christian New Testament involving homosexuality is found in Chapter 1 of Romans, the Apostle Paul's letter to the Romans. If one isolates a few of the verses in Romans Chapter 1 and does not use literary integrity, then it is easy to reach the conclusion that gays are very depraved people. However, a careful reading of various references in Romans Chapter 1 sheds more accurate light on things. In Romans 1: 18 - 32, Paul is addressing people who knew God's teaching. He keeps referring to them as **they**. This begins in verse 18 describing a certain type of man or groups of men over centuries and millennia. These are *"men who suppress the truth by their wickedness, since what may be known about*

God is plain to them. " Continuing his thought, Paul begins verse 21 with *"For although **they** knew God, **they** neither glorified him as God nor gave thanks to him, but their thinking became futile and their foolish hearts were darkened."* As Paul continues, he is obviously speaking of the progression of sin in man over time as a result of the suppression of truth by wickedness. Verse 23 mentions how men made idols in the forms of birds and animals and reptiles. In verse 24 he mentions sexual impurity [in general]. Verse 26 and 27 mention that God gave them [men who suppress the truth by their wickedness] over to *"shameful lusts. Even their women exchanged natural relations for unnatural ones. In the same way the men also abandoned natural relations with women and were inflamed with lust for one another. Men committed indecent acts with other men and received in themselves the due penalty for their perversion."*

Today, there are heterosexual men who may or may not express faith in Christ. They may dabble in pornography or have many partners. As these men proceed further into sexual activity, it may lead to *"indecent acts with other men [plural]."* Some may contract HIV or other diseases. This type behavior matches the tone of this passage. It does not appear to describe heterosexuals and homosexuals who are monogamous and generally under control. But for those who want to condemn all gays, it is important to continue following Paul's discussion of the progression of sin, and all the abhorrent human acts he mentions. Here is the passage immediately following verse 27:

*"Furthermore, since **they** did not think it worthwhile to retain the knowledge of God, he gave them over to a depraved mind, to do what ought not to be done. **They** have become filled with*

*every kind of wickedness, evil, greed, and depravity. **They** are full of envy, murder, strife, deceit and malice. **They** are gossips, slanderers, God-haters, insolent, arrogant and boastful; **they** invent ways of doing evil; **they** disobey their parents; **they** are senseless, faithless, heartless, ruthless. Although **they** know God's righteous decree that those who do such things deserve death, **they** not only continue to do these very things but also approve of those who practice them."*

Of the many sins Paul described in Chapter 1, millions around the world have focused on homosexuality as the unpardonable sin. Why not gossips or slanderers? Thankfully, it does not matter much; because Paul continued his teaching,

"You, therefore, have no excuse, you who pass judgment on someone else, for at whatever point you judge the other, you are condemning yourself, because you who pass judgment do the same things. Now we know that God's judgment against those who do such things is based on truth. So when you, a mere man, pass judgment on them and yet do the same things, do you think you will escape God's judgment? Or do you show contempt for the riches of his kindness, tolerance and patience, not realizing that God's kindness leads you toward repentance?"

(Rom. 2. 1 – 4) A necessary word of instruction: Paul is not saying that each of us has committed each of these sins or to the greatest extent. Paul is simply being consistent with his teaching of the purpose of the Law regarding the Christian faith. Each of us is to recognize that we fall short of perfection. It is where Christianity deviates from many other spiritual ideologies. Self-righteousness has no place.

This is huge. Paul has embarked on decoding the judgment and forgiveness God gives. In other words, Paul is beginning his dissertation on the Christian faith. But too many people do not understand this. This lengthy letter to the Romans – in addition to his other letters in the New Testament - thoroughly explains the connection of the Law to God's plan and to our sin, to God's wrath and judgment, to God's grace and kindness, and to provisions for those who never hear the message. All these nuggets of wisdom, written by perhaps the most qualified scholar to share them, are contained in the 16 chapters of Romans. But tens of millions of today's professing Christians have lost the message early in Chapter 2, because they don't choose literary integrity over peer pressure. And they get lost in self-righteousness.

There are many reasons why it is important to address this issue. Many people of integrity will not even consider Christianity due to the treatment of gays. Prominent, respected evangelists have insisted that Christians confront the LGBTQ community aggressively. Just like Paul rebuked Peter for straying from accurate doctrine, Paul would rebuke many of these prominent men and women of God today for their campaign against people within the LGBTQ community.

Life is complicated. Over-simplifying is dangerous. There is a passage in the Law in the Tanakh commanding that gluttons be put to death. Therefore, if prominent pastors are going after gays aggressively, then why not go after overweight individuals aggressively as well? Thankfully, neither group should be attacked aggressively. The command to put gluttons to death will be discussed in accurate context later in this book, and we have already seen why gays should not be targeted. The message of Christ is grace and dedication among flawed human beings.

It is appropriate to investigate sexual tendencies further. People in the LGBTQ community deserve this level of respect. We should believe the best in one another. Recognizing that Paul was discussing the behavior of humans over time, what role does nature and nurture play in the life of any one person? Applying literary assessment to Chapter 1 of Romans, Paul is not talking about the progression of a specific person over time. Rather, he is discussing the change of humanity over periods of time. Science attempts to quantify how certain behaviors, traits and tendencies can be transferred from a parent to a child. Who are we to assume with certainty that someone who feels a strong predisposition to be gay is not identifying the truth? Additionally, consider behaviors of persons who have been molested. There are many people who have eating disorders or act promiscuously as a result of being assaulted by another person? Given the complexity and uncertainty of anyone's background, most people recognize that it is not our job as Christians to be God's little *sin-finders*. Those types of condemning individuals are scorned by God. Rules and laws and when to evoke them and when to provide grace are discussed in the Tanakh and in the New Testament of the Bible. And the optimal recognition is to appreciate the grace God gives us.

In chapter 3 of his letter to the Galatians, Paul explains that the covenant God established with Abraham preceded the Law by over 400 years. It was a covenant – a promise – made by faith. The Law does not supersede the promise.

"The Law, introduced 430 years later, does not set aside the covenant previously established by God [with Abraham] and thus do away with the promise. For if the inheritance depends on the law, then it no longer depends on the promise; but God in his grace gave it to Abraham through a promise.
Why then was the law given at all? It was added because of

transgressions until the Seed to whom the promise referred had come." (Gal.3.17-19) This will be further discussed in a later chapter. And Jesus is the *Seed from Abraham* - hence, the genealogies in the gospels confirming this. God found a generally righteous and faithful man (Abraham) willing to sacrifice his son. Thus, God could sacrifice his righteous son and seed of Abraham (Jesus) to create the portal to an afterlife. Paul explains in Galatians 3:10 that *"all who rely on the works of the law are under a curse."* In other words, all who believe it is by ending our sin and living righteously that we make it to heaven are cursed. Human life after the death and resurrection of Jesus is often referred to as the "dispensational period of grace." This is what Paul is trying to help us see in his letters relating the covenant to the law and to the act of Jesus. The easiest of all temptations for Christians to fall into may be self-righteousness.

Gandhi was quoted as saying, "I like your Christ, I do not like your Christians. Your Christians are so unlike your Christ." It is Paul's contention – expressed in Romans 2: 18 – 24 - that we all fall short. The law given to Moses was added because of transgressions.

"If you know his will and approve of what is superior because you are instructed by the law; if you are convinced that you are a guide for the blind, a light for those who are in the dark, an instructor of the foolish, a teacher of little children, because you have in the law the embodiment of knowledge and truth – you then, who teach others, do you not teach yourselves? You who preach against stealing, do you steal? You who say that people should not commit adultery, do you commit adultery? [simply looking at another person lustfully equates to adultery.] You who abhor idols, do you rob temples? You who boast in the law,

do you dishonor God by breaking the law? As it is written: 'God's name is blasphemed among the Gentiles because of you.'" (Rom.2.18 – 24)

Ironically, the very source Christians use to condemn others – Paul and his words in Romans – is the very source that condemns those Christians for such an evil twist on Paul's point. There is nothing new under the sun. Self-righteousness and condemnation of others is age-old behavior. To be clear, Paul is not saying that we can all freely sin. [see Romans 3:8] Rather, Paul is encouraging us to carry the accurate message of God. The message is that this life can be brutal. But this life is brutal by design due to the need to prepare our hearts for the promised life afterward. The portal is through the covenant established with Abraham and fulfilled through Jesus. And it is available to ALL people groups. [See errant belief 10 in a few paragraphs].

We Christians are the responsible parties. But we sometimes abuse our power. We have the hope and the message, but too often we use our power to harm and not to encourage. We condemn people into depression and suicide and force them to be opponents of God. We make it easier for them to take a little truth from science and ignore the gaps in some scientific arguments to convince themselves that no God exists; because they would much rather believe in no God than in a God of such insensitive condemnation. And we American Christians need to abandon the dysfunctional behavior of seeing ourselves as the victims. The decline of Christianity in the West is our fault. Many are not preaching an accurate gospel. We should be very concerned. Radical Islam and radical Christianity are typically fueled by self-righteous attitudes coupled with the belief of being victimized. If we allow the misguided condemnation and mistrust to continue, we will see more and more violent acts from radical Christians as well as radical Islamists. If we refuse to fight the rhetoric, we are to some degree responsible for the murder that results.

7 The belief that most American Muslims are in favor of radical terrorism to bring down the West

The recent disrespect and mistrust of most American Muslims is very disturbing. To assume that any religious group is defined by its extremists is illogical, and many people tend to over-simplify what it means to be Muslim. Those of us with Muslim friends and acquaintances find it disheartening that so many accept the stereotyping of all Muslims. Muslims, Jews, Christians along with many other groups agree of our need to battle against Islamic radicals and Christian radicals – especially anyone willing to take another life in an act of terror. We can agree that those groups must be stopped. They murder innocent men, women and children. They do not honor the God of Abraham, and their vision of God's will on Earth is disgraceful to all the Abrahamic faiths and to humanity. Those who condemn all Muslims as extremists provide some of the fuel for the radicalization of individuals around the globe. People can become radicalized toward violent acts in the name of Islam or in the name of Christ.

This is not to say a Christian should accept the teachings of the Muslim faith in their entirety. While there are many common accepted teachings by Christians, Muslims and Jews, there are many differences. Obviously, the view of Ishmael and Isaac are drastically different; and the teaching of grace and rebirth through the life and sacrifice of Christ via the tenets established in the Law given to Moses contrasts Islamic teaching. But people of all faiths should be respected. One can sit and learn from people who hold beliefs that are different. We can disagree and feel passion about the afterlife and how it is achieved. And ultimately it is each of our responsibilities to seek the right path and hold ourselves accountable. If all of us could learn the respect and thirst for truth displayed in the House of Wisdom in Baghdad during the Golden Age of Islam, the world would be dramatically more beautiful and less dangerous.

8 The belief the devil is solely responsible for the presence of pain and suffering and evil
Some might wonder how an educated person can actually believe in a spiritual realm. Frankly, there is no reason to doubt such a realm. Flowing from the previous discussion of God being in the past, present and future [the same yesterday, today and forever], God created every being knowing exactly what he or she would do. Consider the Christian teaching of created angels that turned evil and the teachings of evil principalities and powers in a spiritual realm. God created these and was fully aware of how they would behave. Blaming all pain and suffering on Lucifer and not recognizing that God created him intentionally, once again would suggest that God is not in control. If God is not in control, then we are all in deep trouble. The only way to avoid contradiction (and thus make it possible to find the accurate logical understanding) is to admit that God is not fooled, and he does not make mistakes. Satan was not a mistake. One of the most difficult notions to wrap our brains around is why God would create such evil entities and such a dangerous universe and life. Though he does not target us individually, He allows atrocities, diseases and famines

to threaten humanity. Jesus said, "In this world you will have trouble." (John 16.33) It is the only conclusion that makes sense with a God in control. The question to ask is "Why?". Why so much pain and suffering in this life? It will be discussed later in this book.

9 The belief God is too nice and too loving to create a hell
Wrong. Consider one simple aspect of life on Earth. We (because of our collective selfishness as humankind) and God (because of His purpose for us experiencing this life) allows more than 16,000 little kids to die every day on this planet from disease, lack of clean drinking water and lack of food and medicine. God allows this every day. These children who die – having not yet reached an age of accountability – will be ushered into the afterlife with God. [Consider Paul's argument in the next section.] But consider that God allows these children to suffer greatly in this life. And dying of starvation is painful. Perhaps we should recognize if He does not stop these instances of kids suffering and dying the first death, God will have no problem sitting by and watching adults suffer in the second death [eternity]. The notion that He is too nice to have a hell is illogical, unscriptural and mathematically would defeat a primary purpose of danger in this life. It is through grace that anyone is allowed into heaven. Statistically, given how humans are wired, more people are nudged into seeking and recognizing God's plan via the pain and struggle in this life and with the threat of a hell.
Often, we witness humans at their greatest levels of valor and character during a crisis. We are moved to tears in many cases and may be moved positively in a spiritual direction. God cares deeply for every creature and is moved as well. A loving God

incorporates pain, suffering and evil in this life along with the threat of hell in order to produce the optimal afterlife in multiple dimensions of space and time, and in order to get the greatest number of souls there. A less-loving God would structure this life differently and make it less traumatic. He wants us to enjoy eternity with grateful, rich hearts. This life is necessary to accomplish the next life, where beings with free wills willing to have transformed hearts can interact safely and respectfully and incredibly with new bodies in a realm with more dimensions of time and space. The next life will be mind-blowing. But this train of thought is a brain-bender, no doubt.

10 The belief those outside of Christianity cannot enter heaven
We will confirm the validity of the covenants established through Abraham and Jesus. The signs and plan are quite clear. But Jesus said something that appears to be exclusionary; and it is where many get the notion that no one else will be allowed into heaven other than those who have made a commitment to Jesus in this lifetime. Jesus said, *"I am the way, the truth and the life. No one comes to the father except through me."* (John. 14.6) This seems straightforward and exclusionary, but it is not. Consider a sincere question posed by many conscientious people: "What happens to people in remote parts of the world who never hear about Jesus? Are they going to hell?" Thankfully, Paul gave us the answer in Romans 2: 12 – 16. Again, it is challenging to understand all major concepts surrounding the Law. It is why an entire chapter is devoted to it later. The contrast of the letter of the law versus the spirit of the law very much applies to what Paul teaches.
"All who sin apart from the law will also perish apart from the law, and all who sin under the law, will be judged by the law.

For it is not those who hear the law who are righteous in God's sight, but it is those who obey the law who will be declared righteous. (Indeed, when Gentiles, who do not have the law, do by nature things required by the law, they are a law for themselves, even though they do not have the law, since they show that the requirements of the law are written on their hearts, their consciences also bearing witness, and their thoughts now accusing, now even defending them.)
And just to be clear that Paul is indeed speaking of the judgment and the afterlife, the passage finishes with this: *This will take place on the day when God will judge men's secrets through Jesus Christ, as my gospel declares."* (Rom. 2.16)
He is saying that those who do not hear the gospel will be judged by how they follow the leading of their heart during their lifetime. We will see through the Law and from before time began that the beating and death of God himself – the Son – is the gateway to the afterlife. Just as priests were taught to sprinkle the blood of an atonement sacrifice on the mercy seat in the tabernacle on Earth, the blood of God would be sprinkled on the mercy seat in heaven. It is only through this act of grace that anyone is allowed into the afterlife prepared by God - including the ones who never hear the gospel or the Law. Paul suggests they will be accepted due to faithfully following their heart (and not knowing to put faith in Jesus specifically). With this recognition, Jesus is *"the Way, the Truth and the Life."* (John. 14.6) No one will come to the Father except through him. Some Christians argue vehemently against this, citing Romans 10:9 - 10, where Paul states that we are saved by confessing with our mouth that *"Jesus is Lord."* The verse does not logically imply that confessing Jesus is the <u>only</u> way. After

all, most of us would agree that Moses, David, Isaiah and Daniel will be in heaven. None of them confessed Jesus as Lord. They lived centuries before the crucifixion. They did not know him by name. The covenant with Abraham and the death and Resurrection of Jesus is the portal to the afterlife, and Paul instructs us in Romans 2 that there will be people entering through this portal who did not formally know who Jesus is or what he did.

An important distinction to be made is that the exception of those following their heart does not apply to anyone reading this book. Most persons on the globe today have access to the message of Jesus. God will judge everyone based upon their choices given the knowledge available to them. Remember, there is incredible precision in this universe and in this physical life and incredible precision in this spiritual life. Thankfully then, all people groups are loved and accepted by God. (Isa.42.6 and Luke.2.30 – 32) This life is a brief audition, and the stakes could not be higher.

One final thought – particularly to those who have been negatively affected by Christianity. Paul and Jesus turn religiosity and self-righteousness on their heads. The compassion and love Jesus displayed in his time on Earth reminds us of this love God has for us. Jesus, when accurately defined, is very much who most hope God to be. People who promote an inaccurate definition of Jesus and an inaccurate Christian message can inflict tremendous pain amplified by guilt and shame. This is not a minor issue. On a micro level, it can lead to depression and suicide. On a macro level, it can lead to social injustice and even to genocide. Yet a strange recognition is that a God truly in control is allowing such inaccuracy and such trauma in this life.

If followers of Christ want to see a mighty move of God, these 10 errant beliefs will need to be addressed. Too many non-Christians who generally live with integrity have many reasons not to be attracted to Christianity. The focus will need to shift from much judgment and arrogance to more grace and love in general. In the words of the Apostle Paul to those who engage in judging others so harshly, *"...do you show contempt for the riches of his* [God's] *kindness, tolerance and patience, not realizing that God's kindness leads you toward repentance?"* (Romans 2:4)

Chapter 4

The Fab Five:
Abraham, Moses, David, Paul, and Melchizedek

Context and connections help construct a solid spiritual ideology.

The people through whom this spiritual ideology is established are actual historical figures considered to be significant by the Abrahamic faiths.

Abraham 2000 B.C.

The most interesting man in the world – or at least the most complete. Many think of Abraham as an old man, but in his prime he was a stud. He was extremely successful financially, militarily, and as a respected leader.   Above all, he had integrity. Genesis chapter 14 conveys how he took 318 of his trained men, born in his household, to defeat five evil kings who were terrorizing the region. As is customary, the victors had the right to the spoils. While he allowed his men to retrieve whatever they wanted belonging to those they had conquered, Abraham made it clear he would not take a single item. This man had incredible devotion and faith in God. He wanted everyone to know that his wealth came directly from God's blessing and nothing else.

Jews, Christians, and Muslims all point to Abraham as their

patriarch. This man's man was the one through whom God established His covenant to redeem any individual and allow her or him into the afterlife. God promised a son to Abraham in his old age. After the boy reached a certain age, God asked Abraham to sacrifice the boy to Him. It was a brutal decision that had to be made. Abraham was faithful. He was lifting the knife in the air to complete the request when God told him to stop and, instead, to sacrifice the ram whose head was stuck in a nearby thicket [a sign of the Messiah ultimately having his head in a crown of thorns]. (Gen 22) God had found on Earth a man of character willing to sacrifice his son. Thus, God had the right to sacrifice His son for the sake of mankind. Two thousand years later, the seed from Abraham, his great-great-great … grandson, Jesus, was sacrificed for mankind. And it was done in nearly or the exact location where Abraham was asked to make his sacrifice.

Moses 1500 B.C.
Moses was the one used to deliver God's people from slavery in Egypt. His life intentionally [on the part of God] parallels the life of Jesus. Both escaped death as infants. Moses' mother chose to put Moses in a basket and carefully send it in the river where and when Pharaoh's daughter could find him. Pharaoh's daughter did find him and raised him as her own. Thus, Moses was a Hebrew, of the enslaved nation, but was raised as a prince of Egypt. In his later years, Moses became the deliverer, promised to the Hebrews for centuries. He battled an enormously powerful Egyptian empire and the forces of evil behind it to free many thousands from slavery. (Gen 7 – 12) Jesus, who also escaped death as an infant, would later take on the ultimate spiritual empire of evil and deliver into the afterlife

anyone who accepts his act. Jeremiah prophesied a loss of children around the time of the Messiah, and the Gospel of Matthew records a decree by King Herod to have all young boys put to death in hopes of killing a newborn king born in the area. [Matt 2: 16 – 18] [Note: corroborating evidence for the slaughter of the innocents by Herod depicted in Chapter 2 of Matthew is difficult to find. One of the best sources for information around that time are the writings of Josephus, the Jewish historian of the day. Though he explains in his writings both the early success of Herod and his later murders of family and others, Josephus does not record the killing of boys under age two in this context. It does not mean that it did not happen. It simply means that it is for now unverified. Given the more dramatic executions of Herod's reign, it might have gone virtually unrecorded at the time. And such an action would lack historical significance until more than thirty years later when Jesus was crucified. Therefore, Josephus may not have been privy to such an event. And his mission was the recording of Jewish history and not the recording of the emergence of the Christian church.]

The Law was given to Moses, who recorded it among the first five books contained in the Tanakh. In these five books are key depictions of the creation of the universe that match what today's scientists suggest. In it are descriptions of events and people groups and locations that have helped archaeologists locate sites in recent times. In his message are the ten commandments and other rules that governed a society needing to endure the threats it would face. In these books is the Law, the ridiculously knit-picky set of rules that convinces us of our shortcomings morally. And within the Law and procedures practiced, were the justifications, revelations and prophecies of the plan - the blood payment from God Himself - that would usher humans into the afterlife.

David 1000 B.C.
David was an intriguing person of passion, commitment, bravery, and humility. His life and relationships are the makings of a very complete and compelling novel. In fact, if one claims the Old Testament of the Christian Bible [Tanakh] is boring, that person should read 1st and 2nd Kings and 1st and 2nd Samuel. It describes David's emergence as a chosen boy, his incredible bond with the king's son, and his years of devotion to a king-gone-mad. It describes his bravery and rugged life as a military figure and the years living as a fugitive. Eventually, David became king. But he later committed adultery and had that woman's incredibly dedicated husband put on the front lines of war so that he would be killed. David wore his emotions on his sleeve, took responsibility for his actions, and displayed a child-like devotion to God.

He wrote some of the more critical prophecies of the Messiah, including quoting Jesus on the cross 1000 years ahead of time in Psalm 22: *"My God, my God. Why have you forsaken me?"* The first use of crucifixion according to historians was 800 B.C. Thus, David shared details of crucifixion 200 years before crucifixion was even devised on planet Earth.[1] The recordings of Moses and David are given special significance in Islam as opposed to the way in which other books of the Tanakh are viewed.

Paul [Saul of Tarsus] 1st Century AD
The conversion of Saul of Tarsus – from a zealous Jewish Pharisee and Roman citizen who persecuted early Christians to one who gave up his wealth and stature to travel the world sharing why the Tanakh clearly points to Jesus as the Messiah – is very strong evidence for the resurrection of Jesus. Saul was a

no-nonsense man. He was a Roman citizen and a highly educated Jew. He was devoted to God as he perceived God through the study of the Tanakh and via tradition as a Jewish Pharisee. He originally assumed that Jesus was a fraud. This powerful Saul of Tarsus was present and supported those who stoned to death Stephen, the first Christian martyr. Saul felt early Christians were heretics. He was rounding them up and throwing them in prison. Early followers of Christ were terrified of him. The Book of Acts in the Christian New Testament shares the account of Saul on the road to Damascus to round up more Christians. Jesus had already been crucified and many claimed he had risen from the dead.

"As he [Saul] neared Damascus on his journey, suddenly a light from heaven flashed around him. He fell to the ground and heard a voice say to him,
'Saul, Saul, why do you persecute me?'
'Who are you Lord,' Saul asked.
"I am Jesus, whom you are persecuting.," he replied. 'Now get up and go into the city and you will be told what you must do.'"
[Acts 9: 3 - 6 NIV]

It took time for the reticent Christians to accept Saul as one of them when he first converted. But Saul's undeniable devotion confirmed the sincerity of his conversion. He soon became known as the Apostle Paul. It appears part of God's plan to have an intellectual who had wrestled with his own bias and ideology become the most significant and foundational teacher in all of Christianity. It was Paul (formerly Saul of Tarsus) who let the Jewish believers know that scripture declared the Messiah to be a light to the Gentiles as well as to the nation of Israel. (Acts 13.47, Isaiah 49.6)

He claims to have met the risen Christ. He gave up his stature and wealth. He studied the Tanakh intently most of his life. His humility allowed him to question himself and articulate God's plan. And for years, mending tents to pay for his needs, he went throughout the Middle East, Europe, and Asia instructing all who would listen regarding God's plan. He fled for his life on many occasions and was arrested more than once. Ultimately, Paul was beheaded by the Romans for his faith in Jesus Christ. He is uniquely qualified to teach us all.

The first four of the Fab Five

The first four of these five figures were critically important and faithful. They had tremendous devotion toward God. Yet they all made huge mistakes. Abraham tried to help God's promise of a son in old age by producing a son through a maidservant. (Genesis 16) The two people groups – Arabs and Jews – have fought for most of the centuries since. Moses killed a man and fled. (Exod 2. 11 – 15) David committed adultery and plotted to kill the woman's husband. (2 Sam 11) Paul miscalculated who God was, and was responsible for the deaths of many believers. The first four key players do not have clean records. This, too, is by design and further confirmation of how - even the best among us - fall short of God's standard. This awareness alone ought to instruct us to be less judgmental and more humble as believers and followers. But our human nature fights against us. It is so easy to slip into the delusion that we have left wrongdoing far behind us and emerged as nearly infallible.

Melchizedek, the fifth and greatest of the Fab Five

Moses recorded in Genesis (Gen 14.17 – 24) the meeting between Abram (later to be called Abraham) and Melchizedek before the establishment of the covenant [promise] between God and man around 2000 B.C. Melchizedek was a high priest and king of Salem. He blessed Abram, Abram gave him a tithe, and then the two broke bread and shared wine. Recall how much devotion and integrity Abram displayed toward God. It is enlightening that Abram gave Melchizedek a tithe – presumably one tenth of his wealth. This Melchizedek is very significant. David claimed that the Messiah would be a priest in the order of Melchizedek (Psalm 110.4). The writer to the Hebrews in the Christian New Testament joins David in declaring Jesus – the Messiah – as being a priest forever in the order of Melchizedek – not an earthly priest. In Hebrews Chapter 7, verse 3, the writer proceeds to give us the final insight into the identity of Melchizedek. He was *"without father or mother, without genealogy, without beginning of days or end of life, like the Son of God, he remains a priest forever."* This is not a created being. This is someone who has no beginning or end. This implies Melchizedek exists beyond our space-time dimension and is part of the godhead – the Son, Jesus. Melchizedek is Jesus. Recall the idea that God is one God with three distinct persons in tandem. Moses recorded God stating He would *"make man in our image."* Moses also met this God who referred to Himself as I AM. In John's New Testament Gospel, he records Jesus nearly being stoned to death. Jesus said in response to a faithless comment, *"I tell you the truth, before Abraham was, I AM!"* Abraham broke bread and had wine with Melchizedek [Jesus] just prior to God establishing the covenant with man through the willingness of Abraham to sacrifice his son. Two thousand years later, Jesus [Melchizedek] broke bread and shared wine prior to the fulfillment of the covenant with mankind.

Abraham and Paul met Melchizedek before and after his human life, respectively. Moses wrote signs and prophecies about Melchizedek into the Law, and David recorded the death of Melchizedek - the Messiah, Jesus – 1000 years before it happened.

The Fab Five are connected via Melchizedek [Jesus] before and after his time as a man. The four men lived hundreds of years apart, yet there is no doubt of God's impact in their lives and the significant roles they play in outlining and pointing to God's plan and to God's grace... *given to us before time began.*

Chapter 5

Time and the Holy Book

*How can God be in control but give humans a free will? And when God referred to himself as I AM, why was it a very concise and accurate description? Also, **string theory** does not rule out multiple dimensions of time.*

Paul gives us a scientifically accurate notion of our time in this universe. In 2 Timothy 1:9, he suggests that this grace in Jesus was given to us before time began. Scientifically and from a point of creation, our space-time dimension began with a singular, big bang. Paul is letting us know that God existed before our time began. In other words, God exists in and beyond our space-time dimension. It is impressive that someone in the first century A.D. would express such a notion. Visualizing dimensions helps us better understand the implications of multiple dimensions of time. Below are three figures. The first is one-dimensional. It is a line. It has length, and that is all. The second is two dimensional. It is like a rectangle. It has length and width. We measure its area to denote its size - such as 24 square feet. The third is three dimensional. It has length, width and depth. It has surface area on all 6 faces, but its interior contains space or volume. It is measured in cubic units like a freezer that has 24 cubic feet of volume. Considering these as dimensions of time rather than space allows us to ponder God's relationship to us.

Time line

Time Area

Time Space

The scientific nature of our universe suggests activity in nine dimensions of space within the first split second of the universe.[1] Each dimension is perpendicular to the other eight dimensions. Good luck visualizing that. But consider that a Creator exists beyond our space-time dimension. The implication is that this creator exists in at least one more dimension of space and at least one more dimension of time than we have in our universe. Him existing in at least one more dimension of time helps to reconcile many questions we might have regarding spiritual matters.

Read the remainder of this chapter very carefully. It explains how God has infinite time in any one moment of our time from which he can work. It also discusses how God exists all at once in our past, our present, and in our future. It helps us grasp how he has control while giving us a free will, because he knows every action anyone will ever make – good and bad. Finally, it helps us recognize spiritual beings can exist in another dimension of space; and yet, God ultimately knows every move they will ever make. It's like playing a game where God sees everyone's actions in the past, present, and future. He is ultimately in control of the game and destined to win.

There are various passages in scripture and in the New Testament that speak of another dimension of time.

Paul suggests that God existed *before our time began*. In the letter to the Hebrews in the New Testament, the writer [unsure if it was Paul or someone else] claims that *"Jesus is the same yesterday, today and forever."* (Hebrews.13.8) Recall that this same author wrote about Melchizedek (Jesus), that he had *"no beginning of days and no end of days."* [Hebrews 7:3] These are all descriptions of God existing in at least one more dimension of time. Of additional interest is a prayer Jesus made to God, the Father recorded in John 17: 4, 5: *"I have brought you glory on earth by completing the work you gave me to do. And now, Father, glorify me in your presence with the glory I had with you before the world began."*

Consider the rectangle below – like a sheet of paper – to be two-dimensional time. Let the line – like a string – represent the timeline of the universe. The string is sitting on the rectangle. The left end of the string is the beginning of our universe – the beginning of our time. The right end of the string is our current time extending into the future. The string will continue to increase slowly to the right as time proceeds. The rectangle gives an idea of what God can do if he exists in a second dimension of time. In this scenario, he is present anywhere on the rectangle simultaneously.

God's Presence

Timeline of Universe

God's Presence

God's Presence

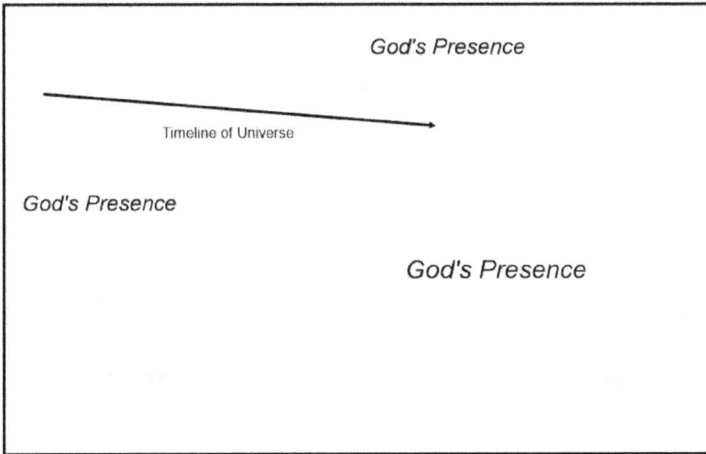

And from the perspective of our timeline, he is at all times at once – *the same yesterday, today and forever.* At any one moment of time along the timeline of our universe, God exists there. He can operate in the lives of 7 billion people simultaneously. And He simultaneously exists in the past and in the future. This is how God can impress upon David to write a Psalm describing Jesus on a cross 1000 years before it occurred. At this very moment in our lives, God is here with us, back at the cross with Jesus and back in 1000 B.C. with David. Nothing surprises him, allowing Him to be in control while giving people a free will. This is likely why God described Himself to Moses very concisely as I AM. God just is… everywhere…simultaneously. I AM.

Jesus is one of three persons of God and was nearly stoned to death by religious leaders when he said, *"Before Abraham was born, I AM."* [John 8:58] In fact, looking at the few verses preceding his declaration, we are brought full circle back to Jesus [Melchizedek] with Abraham 4100 years ago. Here is the dialogue recorded between Jesus and some of the biased religious leaders of his day:

Jesus:
*"'Your father Abraham rejoiced at the thought of seeing my day,
he saw it and was glad.'*
*'You are not yet fifty years old,' the Jews said to him, 'and you
have seen Abraham!'*
*'I tell you the truth,' Jesus answered, 'before Abraham was born,
I AM!'*
*At this, they picked up stones to stone him, but Jesus hid himself,
slipping away from them.'"*
God being I AM is ultimately in control in our present universe.
Nothing can separate us from Him – not even time. Here is a
passage from Paul's letter to the Romans.
*"For I am convinced that neither death, nor life, neither angels
nor demons, neither the present nor the future, nor any powers,
neither height nor depth, nor anything in all creation, will be
able to separate us from the love of God that is in Christ Jesus
our Lord."* [Romans 8:38,39]
In Luke's Gospel of Jesus Chapter 24:44, he records the words of
Jesus after he had risen from the dead.
*"This is what I told you while I was still with you: Everything
must be fulfilled that is written about me in the Law of Moses,
the Prophets and the Psalms [Writings]."* The Law, the Prophets
and the Writings [Psalms] are the Jewish Tanakh.
There are statements in the Tanakh (Old Testament of The
Bible), that we know are scientifically accurate today but should
not have been known thousands of years ago, when they were
recorded. These scriptural passages support the existence of a
being from beyond our space-time dimension influencing the
writers. For example, if an ancient writer was going to describe
how God formed the universe, one would expect a description
of how God *stretched* out the universe – past tense. The

recognition of an expanding universe is a fairly recent scientific discovery. However, many passages in the Tanakh claim that God *stretches* out the universe – present tense. Today, we know this to be scientifically valid. Isaiah 40:22 is a very accurate depiction of our universe today. "He sits enthroned above the circle of the earth, ... He *stretches* out the heavens like a canopy, and spreads them out like a tent to live in." Since God is at all times present, such passages are meant for the modern world as an indicator of His influence on the writing of the Tanakh and in our lives. Such statements in addition to clear descriptors of events that later took place in the life of the Messiah, were written to let future generations know that God influenced the writers of the Tanakh. As important as the Tanakh was to the hundreds of thousands of Jews before Christ, it is much more crucial to the billions of souls on the planet today. Other passages referencing the expansion of the universe include Psalm 104: 2, and Zechariah 12: 1. Job 9: 8, 9: *"He alone stretches out the heavens and treads on the waves of the sea. He is the maker of the Bear and Orion, the Pleiades and the constellations of the south."* Today we realize that these constellations move further apart as part of a universe that is not static – it is expanding. It indeed *stretches* out. One other point worth mentioning from the 9th verse of Job 9. It says that God alone treads on the waves of the sea. Recall that Jesus was reported to have walked on water. (John 6: 16 – 24) This also identifies him as God.

The Genesis account describes a beginning of time – which is unusual in holy books – while being scientifically accurate.[2] And if one reads the passage with literary integrity, it is in line with today's model of the evolution of our universe from 13.7 years ago until humans emerge. Genesis 1:2 gives us the perspective from which to visualize the passage. *"The Spirit of God was hovering over the waters."* (Gen.1.2) Reading the passage from the perspective of being just above the surface of the planet, causes the scriptural account to line up with a scientific model of how the Earth evolved. The combination of verbs used in Chapter 1 of Genesis implies that the heavens and Earth were created brand new out of nothing seen – *"In the beginning, God **created** the heavens and the earth."* (Gen.1.1)[3] But a different verb combination is used to describe the emergence of the sun.[4] It is why translators of the original Hebrew chose the words, *"**Let there be** light,"* (Gen.1.3) as opposed to God *created* the light. The Hebrew text translated *let there be light* is conveying that the light was already in existence but is now noticeable.[5] This is because the Earth was initially covered in dense cloud cover like Venus – which the Book of Job correctly describes. (Job 38.9) Early on there would be day and night, but no visible view of the sun or moon. It was during the Hadean Period of asteroid strikes that the dense cloud cover was broken up, and the sun was finally visible. Thus, the writer of Genesis records on day four – the fourth long period of time – that the sun was finally visible. *"**Let there be** lights in the expanse of the sky"* (Gen.1.14) Once again the author chose not to say that the light was *created* at that point in time. Instead, the verbs imply that the sun and the moon and the stars existed already but now were literally being brought into view from the surface of planet Earth. The precise choice of verbs by the writer matches today's scientific model of cloud cover, the development of an atmosphere and water cycle as the cloud cover was broken up, and then the sun eventually becoming visible.[6] This is remarkable for a book written 3500 years ago.

Chapter 6
The Law: Misunderstood but Vital

The Law and what it helps us understand is challenging and revealing. But if not perceived accurately, it can become a stumbling block for individuals and cultures.

The Law given to Moses is contained in the Torah, the first five books of the Jewish Tanakh and Christian Old Testament. Ironically, the Law given to Moses in 1500 BC is foundational to our relationship with God and our acceptance into heaven; yet, it is often avoided or misrepresented in churches. The Tanakh [Old Testament] is referred to as The Law, The Writings, and The Prophets. Once again humility and integrity are required to put the Law into proper perspective. Some people use it as a club to rule over others. Some use it to justify themselves. Others assume it is so outdated and harsh that it can be ignored. But the Law contains the power to transform us, and it begins by letting us know how far short we fall when using it as the standard by which to judge ourselves. Paul, the ultimate scholar, walks us through this very important portion of the Tanakh. It is wise to read the entire Letter to the Romans by Paul in the Christian New Testament. It is a dissertation on righteousness achieved through Jesus and the promise made to Abraham in connection to the Law. It is an explanation of how Jesus is the fulfillment of the law. Ultimately, it is faith in Jesus – discovered through humility and integrity – that ushers us into the next life.

In Romans Chapter 4, Paul reminds us that Abraham was not justified through the law. The law came hundreds of years later. Abraham was justified by faith. In Romans 4:15, Paul states that the *"... law brings wrath."* In Chapter 7, Paul shares a passage that reveals one of the main purposes of the law. *"What shall we say then? Is the law sin? Certainly not! Indeed I would not have known what sin was except through the law. For I would not have known what coveting really was if the law had not said, 'Do not covet.' But sin, seizing the opportunity afforded by the commandment, produced in me every kind of covetous desire. For apart from the law, sin is dead. Once I was alive apart from law; but when the commandment came, sin sprang to life and I died. I found that the very commandment that was intended to bring life, actually brought death. So then, the law is holy, and the commandment is holy, righteous, and good.*
Did that which is good, then, become death to me? By no means! But in order that sin might be recognized as sin, it produced death in me through what was good, so that through the commandment sin might become utterly sinful." [Romans 7: 7 – 13]

Therefore, the Law helps us to recognize our shortcomings. When initially revealed by Moses, the Law helped the Jews of Moses day to live a strict life that would allow them to survive and carry on. They had to survive against the threats of other cultures infiltrating or annihilating them. In addition, the Law quantified the meaning and ritual of sacrifice, ultimately pointing to the sacrifice of God the Son, Jesus.

As Jesus was nearing death on the cross, Matthew, the converted tax collector, records a very strange statement made by Jesus. Jesus said, *"My God, my God. Why have you forsaken me?"* [Matthew 27: 46] David recorded this moment of Jesus dying on the cross 1000 years before it actually happened. His Psalm 22 begins, *"My God, my God. Why have you forsaken me?"* The Psalm shares many details of the crucifixion. *"A band of evil men has encircled me, they have pierced my hands and my feet. I can count my bones; people stare and gloat over me. They divide my garments and cast lots for my clothing."* This was the scene during the capture, trial, and crucifixion of Jesus. More on this in Chapter 9 on Signs and Prophecies. On another note, when David mentions that his hands and feet were pierced, he is referencing a form of capital punishment – crucifixion - that would not be devised on Earth for another 200 years.[1]

God let the other four of the Fab Five know that the Son [Melchizedek, Jesus] was to be crucified. In 2000 B.C. as Abraham and his young son were walking toward the place where the sacrifices were made, his son asked him, *"'The fire and the wood are here, but where is the lamb for the burnt offering?'*
Abraham answered, 'God himself will provide the lamb.'" (Gen.22.8)

With Moses in Egypt in 1500 B.C., God instructed the Hebrews to take blood from the Passover lamb and put it on the sides and tops of their doorframes to prevent death. (Exod.12. 7) This was a sign of the true Passover lamb, Jesus, who would have his blood put on the wood of the cross to take away the sins of mankind and create the portal to the afterlife. [How many of us realized that the Jews were crossing themselves long before the Catholics took up the gesture?] According to the law, the Passover lamb selected was to be without blemish; and none of its bones were to be broken. This is why David in 1000

B.C., when describing the future crucifixion of the Passover lamb, mentioned in Psalm 22 that *"I can count my bones."* It is also why John, in Chapter 19 of his gospel, recorded that the soldiers needed to get the three men down from their crosses, because the special Sabbath of the Passover celebration would begin at sundown. Breaking the legs of those on the cross prevented them from pushing up to get air. Once the legs were broken, the lungs would fill with water, and death would occur. John records how they broke the legs of the two thieves, but that Jesus was already dead; therefore, they did not break his legs. This is the fulfillment of prophecy recorded by both Moses and David. Additionally, John recorded that they pierced the side of Jesus to confirm that his lungs had filled with water and that he was dead. This was a fulfillment of the prophecy of Zechariah 12:10.

Returning to the law in Exodus Chapter 12, Moses instructed that in future honor and celebration of the Passover in Egypt, the Passover lamb was to be selected on the 10th day of Passover, sacrificed at twilight on the 14th day, and the 15th day was to be a special Sabbath celebration. Recall what John recorded regarding the death of Jesus. They had to get the three off the crosses, because the next day was the special sabbath – day 15. Day was ending. Twilight had begun. It was the 14th day of Passover when Jesus was crucified [sacrificed] at twilight. The next day was the special Sabbath, the 15th day of Passover. Additionally, on the 10th day of Passover, Jesus rode into Jerusalem on a donkey [prophesied by Zechariah in Zechariah 9:9]. All of this was a perfect fulfillment of prophecies from 2000 B.C. through 150 B.C.

The way the events played out in the final week of the human life of Jesus is mathematically profound. For 1500 years, Jewish culture practiced for the ultimate entry of Jesus into Jerusalem. Moses instructed the Israelites that each year would begin with Passover. On the tenth day, each man was to select a Passover lamb. According to John's gospel, Jesus was near Jerusalem in Bethany six days before the Passover feast, which occurs on the 15th day. (John 12: 1) In John 12: 12, 13 the account continues. *"The next day [this would be the 10th day of Passover] the great crowd that had come for the Feast heard that Jesus was on his way to Jerusalem. They took palm branches and went out to meet him, shouting, 'Hosanna [God save us]! Blessed is he who comes in the name of the Lord!"* Some historians have questioned whether it would have been palm branches that were held. Others have suggested that Jewish tradition had the high priest going out of the city to select the Passover lamb, and upon his return, the congregation would wave palm branches and shout *"Hosanna!"* After the death and resurrection of Jesus, John and the disciples read accounts in the Tanakh of what the disciples witnessed regarding the Messiah's last days. In addition to seeing Jesus alive again, the most significant holy book gave many vivid details of what they witnessed. The disciples were reading a story that they watched play out, knowing that the story was written hundreds of years earlier. The impressive prophecy regarding the entrance into Jerusalem is found in Psalm 118. In order to grasp a greater understanding, one can reference another portion of the Law involving cities of refuge. Numbers 35 describes how the Law established six cities of refuge. If a man accidentally killed another man, he had to outrun his pursuers to a city of refuge and plead his case. If it was determined that the death was accidental and not premeditated murder, the man had to remain in the city until a very significant event occurred. The event was the death of the high priest. Once the high priest died, the man was forgiven and free. Obviously, this is a sign of the death of the Messiah [Jesus, Melchizedek – the high priest]. Of further significance are the horns of the altar, where sacrifices were performed. Combining implications recorded in Exodus 21: 12 – 14 and 1 Kings 1: 50 – 52, men ran to the altar and grabbed the horns of the altar to plead their case. It is where they would either be condemned or freed by the death of the high priest. With this background, we can now read the prophecy in Psalm 118 describing the entrance of the Messiah into Jerusalem. First, Psalm 118: 22 states, *"The stone that the builders rejected has become the corner stone."* The Messiah was to be rejected by the leaders of his day, yet he would become the foundation of God's kingdom for eternity. From there the Psalm declares what an important time it is; and it mentions that people would be yelling *"Save us!"* and *"Blessed is he who comes in the name of the Lord."* And then Psalm 118: 27 describes the scene 1000 years after the Psalm was written. *"With boughs in hand, join in the festal procession up to the horns of the altar."* Referencing the horns of the altar ties the event to the death of the high priest in the cities of refuge. The prophecy did not specify that people would wave palm branches. It just stated that the boughs of some sort of tree would be in the hands of those escorting the Messiah into the city in celebration and preparation for the feast. Again, here is what the disciple John recorded:

"The next day [this would be the 10th day of Passover] the great crowd that had come for the Feast heard that Jesus was on his way to Jerusalem. They took palm branches and went out to meet him, shouting, 'Hosanna! [God save us] Blessed is he who comes in the name of the Lord!" That day, neither the

crowd nor the disciples realized what would happen four days later. Four days later would be the completion of the promise to Abraham. Four days later, God's Passover lamb – Jesus - would be sacrificed. Six days later, in heaven – where the actual tabernacle and altar not made by human hands resides (Exod 25: 9, 40; Heb 8: 1 - 6), the high priest Melchizedek – Jesus – sprinkled the blood of God to open the portal to the afterlife for any man or woman whose heart chooses to go there. And then the risen King – Jesus – returned to Earth to reveal himself to his followers and later to Saul of Tarsus. All these events began with the procession into the city on the 10th day of Passover. Isaiah wrote a lengthy description of the life, death and resurrection of the Messiah in Isaiah 53 (circa 700 B.C.).

"He was oppressed and afflicted, yet he opened not his mouth; he was led like a lamb to the slaughter, and as a sheep before her shearers is silent, so he did not open his mouth.

By oppression and judgment he was taken away. And who can speak of his descendants? For he was cut off from the land of the living; for the transgression of my people he was stricken.

He was assigned a grave with the wicked [he was crucified with two thieves], and with the rich in his death (he was buried in a rich man's tomb), though he had done no violence, nor was any deceit in his mouth.

Yet it was the Lord's will to crush him and cause him to suffer, and though the Lord makes his life a guilt offering, he will see his offspring [spiritually] and prolong his days [he will rise from the dead], and the will of the Lord will prosper in his hand. After the suffering of his soul, he will see the light of life and be satisfied [he will return to his spiritual form]; by his knowledge my righteous servant will justify many, and he will bear their iniquities.

Therefore, I will give him a portion with the great, and he will divide the spoils with the strong, because he poured out his life unto death, and was numbered with the transgressors. For he bore the sin of many and made intercession for the transgressors."

The Apostle John recorded the following after the resurrection in John 12: 16: *"At first his disciples did not understand all this. Only after Jesus was glorified did they realize that these things had been written about him and that they had done these things to him."* These were fisherman and lay people. They were not educated like Saul of Tarsus. They must have felt great faith and amazement as they later read the various prophecies in the Tanakh. They lived what they were reading.

There is another interesting detail from Leviticus 19: 5 – 9. The practice of the burnt Passover offering was to sacrifice it one day and eat it on that day or the next. But Moses warned that anyone who ate it on the third day was desecrating what is holy. This is likely because Jesus, the ultimate sacrifice, would rise from the dead on the third day.

Paul explains how God precisely established that Jesus [Melchizedek] could bear the transgressions of us all, and thus establish the portal through which we are able to accept and thrive in an afterlife. God included a specific provision in the Law recorded in Deuteronomy 21: 22 and 23. *"If a man guilty of a capital offense is put to death and his body is hung on a tree, you must not leave his body on a tree overnight. Be sure to bury him that same day, because anyone who is hung on a tree [other interpretations of the Hebrew describe a pole] is under God's curse."*

This is eventually how God was able to curse Jesus on a cross with the sins of mankind - and why Jesus would then utter the words, *"My God, my God. Why have you forsaken me?"* Again, Paul beautifully ties it all together; and he puts the law into perspective at the same time. From his letter to the Galatians, Chapter 3:

"You foolish Galatians! Who has bewitched you? Before your very eyes Jesus Christ was clearly portrayed as crucified. I would like to learn just one thing from you: Did you receive the Spirit [the Holy Spirit, the third person of the Godhead] by observing the law or by believing what you heard? Are you so foolish? After beginning with the Spirit, are you now trying to attain your goal by human effort? Have you suffered so much for nothing – if it really was for nothing? Does God give you his Spirit and work miracles among you because you observe the law, or because you believe what you heard?

Consider Abraham: 'He believed God, and it was credited to Him as righteousness.' Understand, then, that those who believe are children of Abraham. The Scripture foresaw that God would justify the Gentiles by faith and announced the gospel in advance to Abraham: All nations will be blessed through you. So those who have faith, are blessed along with Abraham, the man of faith.

All who rely on observing the law are under a curse, for it is written: 'Cursed is everyone who does not continue to do everything written in the Book of the Law.' Clearly no one is justified before God by the law, because, 'The righteous will live by faith.' The law is not based on faith; on the contrary, 'The man who does these things, will live by them.' Christ redeemed us from the curse of the law, by becoming a curse for us, for it is written, 'Cursed is anyone who is hung on a tree.' He redeemed us in order that the blessing given to Abraham might come to the Gentiles through Christ Jesus, so that by faith we might receive the promise of the Spirit.

Brothers, let me make an example from everyday life. Just as no one can set aside or add to a human covenant that has been duly established, so it is in this case. The promises were spoken to Abraham and to his seed. The Scripture [Tanakh] does not say 'and to his seeds', meaning many people, but 'and to your seed,' meaning one person, who is Christ. What I mean is this: The Law introduced 430 years later, does not set aside the covenant previously established by God and thus do away with the promise. For if the inheritance depends on the law, then it no longer depends on a promise; but God in his grace gave it to Abraham through a promise.

What then was the purpose of the law? It was added because of transgression until the Seed to whom the promise referred had come. The law was put into effect through angels by a mediator. A mediator, however, does not represent just one party; but God is one.

Is the law, therefore, opposed to the promises of God? Absolutely not! For if a law had been given that could impart life, then righteousness would certainly have come from the law. But the scripture declares that the whole world is a prisoner of sin, so that what was promised, being given through faith in Jesus Christ, might be given to those who believe.

Before this faith came, we were held prisoners by the law, locked up until faith should be revealed. So the law was put in charge to lead us to Christ that we might be justified by faith. Now that faith has come, we are no longer under the supervision of the law.

You are all sons of God through faith in Christ Jesus, for all of you who were baptized into Christ, have clothed yourselves with Christ. There is neither Jew nor Greek, slave nor free, male nor female, for you are all one in Christ Jesus. If you belong to Christ, then you are Abraham's seed, and heirs according to the promise."

Paul was one of the first intellectuals to challenge his bias, confront his ideology, hold himself accountable to Scripture [the Tanakh] and to reality, and uncover the ultimate mystery and plan created by God. And information like Paul recognized and like is shared in this book is often lacking in churches today. But today, more than ever, we need intellectuals to recognize what Paul is sharing and to become part of the body of Christ and help to confront and change some damning group-thinks within the religious community as well as within the scientific community. There is one path through the portal. It is precise, like God's universe and this life is precise. And yet, God in his grace, mercy and practical love, will allow anyone into heaven that has not heard of this Jesus and of this plan provided they follow the leading of their heart — the law written on their heart as Paul suggests at the end of Romans Chapter 2. But even they will make it to the afterlife only through the portal God established: the sacrifice of Jesus, God the Son. Therefore, Jesus is indeed, *"The way, the truth, and the life."* No one will come to the Father except through him and his act.

The law of Moses is often misunderstood. To get too hung up on the rules in the Law is to misunderstand its purpose and the manner in which we are to live - namely that we are to live by faith and the love and works displayed and produced from that faith.

Some atheists have thought that they logically destroyed Judeo-Christian ideas by "exposing" crazy ideas in the law. For example, some have referenced children being put to death if they are drunkards and gluttons. (Deut 21.18 – 21) Atheists are horrified that God would command people to kill their children. However, such an argument ignores context. Ego, bias, and self-centeredness [EBS] can make intelligent people reach erroneous conclusions. A simple understanding of the culture and age of accountability in the time of Moses helps us gain a more educated understanding of the situation. Unlike our nuclear families of today's industrial and technological age, in Moses day it was a patriarchal, agricultural society. Grandpa was in charge. If his 24 year-old son was defying what it took to preserve the tribe and was a drunk and glutton running perhaps with the women of the enemy and not at all willing to change, he was to be put to death. Moses was not talking about putting six year-olds to death. He was establishing strict laws to preserve the Jewish race during a more tenuous time.

Chapter 7

The Not-so-Fab Five:
Pain, Suffering, Evil, Guilt, and Fear

This chapter is primarily for the large portion of the population who struggle with or are affected by these five forces.

Pain and suffering are delicate subjects to address. How can God sit back and allow some of the things that are done to people? It would make it seem as though God is unloving or does not exist. It may help a little that he was willing to be born a human – a poor human. And it may help to know that he was willing to be severely beaten and then brutally killed. But if he is God, he had the choice to make things so different. So why didn't he? The answer only makes sense when we compare this short life to an afterlife that will be rich in relationships, creativity, and fulfillment - and will be free of an evil presence because of what is purged via this first life. Compared to billions of years and more in an afterlife that just is, the life we live here is extremely short. The pain of this short life may be compared to the brief pain of receiving a shot of medicine that provides years of good health. And the many joys of this life give us a taste of the life to come.

Some people have had unspeakable things done to them. God did not create those individuals to be targeted, but he certainly

71

did allow it. And it is not scripturally accurate to say, "Everything happens for a reason." Yes, generally pain and suffering are part of this life for various reasons. But individually, this is an inaccurate and cruel statement to share with someone going through a crisis. The teachings of Paul [Romans 8:28 in particular] suggest that *"In all things, God works for the good of those who love him."* And Jesus said, *"In this life you will have tribulations. But be of good cheer. I have overcome the world."* (John 16.33) When life gets difficult, some of the most valuable people available are those who went through a crisis successfully. Often those with beautiful hearts were once subjected to things we would not want to endure. God may have used their pain to increase empathy and develop keen understanding.

First and foremost, we exist with the ability to make decisions. We decide where we place our allegiance, and we constantly make decisions regarding our relationships and interactions with others. Where we each choose to go in the afterlife is also based on the decisions we make. It takes humility and a conscious effort to seek and find spiritual answers. And it requires integrity. Those of us who struggle well through this life and come to the realization of what is available to us – through no great act we have committed but rather through God subjecting himself to extreme pain and bloodshed on our behalf – will enter the afterlife with a perspective that makes the afterlife greater than we could have imagined. Being free from those individuals on Earth who chose arrogance, selfishness, and exploitation will make possible a relational richness with others that will blow us away.

Pain and trauma can be debilitating, but it can also produce some of the most desirable human traits and interactions. In a strange but intentional way, people on Earth who go through a tragedy or trauma with another that they trust, experience a depth of relationship that others may not. Many world relief workers make this observation. They have witnessed immense valor, courage, and goodness from persons who have been devastated, but who selflessly attended to the needs of others. Consider also military personnel who survive trauma and challenges with others in the armed services and build a bond that otherwise would not exist. Unfortunately, the flip side is that they may have to endure lingering pain and trauma that few of us will experience. The thing we often hear from our military men and women returning from conflict is their frustration with those who tend to take so much for granted. It is an important observation. When our lives are too easy, we become lax in many respects. This may be why it is easier for those in poverty to pursue a spiritual journey compared with some who are wealthy.

A note to those in the military who suffer from the effects of what they have witnessed: *such suffering does not make you weak. You have experienced things that the rest of us pray we never do. Sometimes the platitudes or ideas we had implanted in us growing up do not match the reality of life or of war. In such cases, abandon those platitudes and ideas, because God is the God of reality. We can hate Him at times for creating a world with the horrors of war; but the horrors of war are part of reality. Yell at God for what you experienced. You should not be forced into a paradigm that ignores what you endure. Pain does not permanently leave us in this life. In a moment's notice, we can feel grief that is nearly unbearable. The only things that make sense IF there is a God in control and given the brutality of* *this life, is that God knew it would happen and did nothing to stop it. So yell, question, scream, and grieve...until it starts to sink in that – in light of an afterlife – this life is precisely what is necessary to optimize the next life. This life – that is sometimes heaven-on-Earth and sometimes hell-on-Earth – must work its purpose to pave the way for an afterlife of billions of free-willed beings that can flourish in a realm orchestrated by love and trust.*

You are not at all weak for the pain you endure on behalf of the rest of us who reap the rewards of your sacrifice. My prayer for you is that you recognize your humanness – the good and the bad within us all – and cling to your virtue and character even when you may think it has left you. Hang on during the dark moments and be determined to move forward. Get help and encouragement from others. You are not weak in doing so. You are wise. Do it for you, for your loved ones, and for the other members of the military who need your example and encouragement. It is truly a struggle and sometimes not at all pretty. And yet, if we struggle well, it can be a beautiful part of our life. May God watch over you. Thank you so much for what you have accomplished for so many. And may God work through wise, effective individuals and organizations to better care for military personnel harmed by their service. Too often it seems our military personnel are injured twice. Once during military action and again if government agencies turn their backs and fail to take responsibility for some health problems caused by military decisions or practices.

God created the universe exactly the way he intended. He made sure that nature would point to him, whether some can break away from their bias and ideology long enough to consider it. He provided a holy book that is accepted still today by half of the world's population. He provided signs and prophecies through writings and actual life events on the part of many key individuals. And he himself is the person 2000 years ago that so many still point to today as the embodiment of love. The plan is clear and substantiated for those who make an earnest attempt to find it. Ultimately, it is our decision. Will we choose to accept the incredible gift from God that cannot be earned, or we will choose an ideology that fails to acknowledge the purpose of this short life? God wants the greatest number of people to choose the gift. It is this desire that necessitates the proper balance of evil, pain and suffering here. Mathematically, it provides for the greatest number of souls to accept the gift. A lack of evil, pain and suffering (that might otherwise catch our attention) would be evidence of a God who does not care enough to get the greatest number of people to choose correctly.

Unfortunately, the free will, insecurity, self-centeredness and ego we humans possess can unleash horror on others and can subtly form a perspective on the part of its recipients that can haunt and harm them throughout life. Perhaps nowhere is this more true than the impact a parent can have on a child. A parent can desire to do everything right and want the best possible life for his or her child. But parents live with their own stresses and past hurts as they inevitably cause harm in various circumstances throughout the life of their children. Even the best parents have regrets. But they also have a way of training children to recognize that bad moments will occur. Forgiveness and understanding in these experiences can help us all to interact better and to know how to take proper responsibility for our actions. But in cases where parents have become self-absorbed and inhumane, their abuse of their vulnerable children will create a living hell those children will try to overcome for the rest of their lives. And when a parent – who is the natural protector – is the very one to betray a child, the difficulty for that child at any age in life to put trust in others to help them through the hurt is a second assault on their livelihood.

For the many who live with deep hurt from such circumstances, please consider this. The way out is always through a perspective of wisdom and understanding. It is the mind and the heart that ultimately must come together. It is a war that must be waged against the inaccuracies of what our minds or emotions may tell us. Unfortunately, or fortunately, there are many who – though the experiences are always unique – can relate to the injustice and to the battle that must be waged. Not every question has an available answer. But enough questions can be answered; and we must build our perspectives on reliable information to climb slowly out of the darkness and hurt that sometimes feels beyond bearable. Seek the help of others. Love yourself unconditionally and hang on to hope.

A study of the promises in the Tanakh and the recordings of things Jesus said and did help us to recognize that Jesus is part human and part God. And he provides a tangible vision of who God is. Though many justifiably are frustrated with certain Christian behavior today and over the centuries, few would ever question the motives, wisdom, justice and goodness of Jesus. He destroyed religiosity and connected with people in the very way in which they needed. He hung out with those that others rejected. And he had a way – a godly way – of saying or doing the ideal thing that helped open eyes to reality. He ultimately conveyed the wisdom and the understanding that powerfully put things into perspective and permanently changed lives for the better: for eternity. When our perspective changes to an accurate reality, then we can move forward in whatever circumstance and with whatever time we have left to live in a way that makes sense. If we are wise and undaunted, our allegiance then becomes to this Jesus and to His ways. It is not necessarily to the religious ways that all too often deviate from the example and teaching of Christ. It is upon Jesus and the God that understands pain (and allowed it) that we cast ourselves. We can yell at Him, question Him, and wrestle with Him. He can comfort us and give us hope. And our pain can be used to help others. It is similar to the message in the movie, *Collateral Beauty*. Such an idea, that the hardship some of us go through can lead to beautiful support and benefit to others, makes sense to us who have experienced intense pain and attempted to face it, live through it, and let God comfort and work through us. And yet, the pain is never truly gone. It can return in an instant. But if we are willing, it has its place, and it can produce a positive power that other experiences cannot. But it can take time – two steps forward and one step back. Life can be brutal. But it is also beautiful. And Jesus can relate like no one else. We may need to put into perspective how institutions and others affected us and added to our burden. But God and His wisdom is our saving grace. Our allegiance is to integrity and reality and to a practical, enlightened and accountable approach to life.

Before he was captured, Jesus knew the horrible things the Romans would do to his body. He felt it.

"And being in anguish, he prayed more earnestly, and his sweat was like drops of blood falling to the ground." [Luke 22: 44]

The good in many of us wants to suggest that God could never

allow anyone to be hurt. Perhaps we want to believe there is
only heaven and no hell – whatever hell is. Some teach that,
once we accept Christ and follow him, our lives will be without
harm the rest of our days. This does not line up with what we
see in life. All the pleasant thoughts of how we might want this
world to be are not in line with scripture, Paul's teachings or the
reality we witness globally.

When people look at this life without the context of an afterlife,
it makes no sense that a loving God would allow such things.
There would be little reason to serve God if we lived this life,
died, and there was nothing more. Such a life would point to a
cruel God. In light of an afterlife and with a recognition of what
this challenging life accomplishes in preparing the sort of
existence yet to come through the suffering and experience of
this first life, the Plan established in the Tanakh points to the
work of a highly calculated, loving creator. We may not think
we like this world and this life. In fact, none of us always likes it.
But this is reality. And a loving God, who experienced first-hand
a greater brutality than most of us will ever face, is in control.
The chaos and hurt in the world today is not evidence of no God
or of a God not in control. On the contrary, it is what He made
and is exactly what is necessary. He tells us how it is. He let us
know that we will have tribulation. He came down and
experienced it. He promises us an absolutely beautiful afterlife.

When some ponder their mortality, another nudge is provided.
The physician Luke recorded these words from Jesus: *"I tell you,
my friends, do not be afraid of those who kill the body and after
that can do no more. But I will show you whom you should fear:
Fear him who, after the killing of the body, has power to throw
you into hell."* [Luke. 12. 4, 5] Hell is a reality taught by Jesus. If

true, it may be assuring to know that tormentors on Earth will receive a much more horrific torment than they ever inflicted in this life. But there is yet another benefit to the pain and suffering we witness in this life. It is that God sits back and allows it. Consider that on average each day 16,000 children age five and under die of malnutrition, famine, lack of clean water, disease or conflict.[1] Thankfully, this is an improvement since 1990, when 35,000 children died per day on average.[2] Perhaps part of why God allows the deaths of these children is to make a very relevant point. Just as he sits back and allows these children to die in this life, He will sit back and watch people die the second death in hell. We choose to let kids die, and God sits back and allows it. The good news is that these young kids who die – having not yet reached an age of accountability – will join God in the afterlife. The bad news is that, according to the teachings of Jesus and the implication of the Tanakh and the need for God's plan, there is a hell to be experienced by those who avoid the humility and diligence required to recognize, ponder, and accept the purpose and plan of this life.

Prove that there is a hell? We can't. Prove that there is not a hell? We can't. But we can prove beyond a doubt that a God exists and that the plan involving Jesus is the designated passage to the afterlife. And the implications along with the teachings of Jesus and Paul suggest both a heaven and a hell. This life, with its joys and blessings, is also a battle field for souls. Living with understanding, integrity, purpose, love, courage and humility leads to the greatest good on this battle field. The epiphany of God's plan and the humility to grasp it and communicate it is necessary for impacting others positively in the spiritual realm. It is why it is so sad to see the lack of intellectual arguments for the existence of God and for the recognizable plan that paves the way to the afterlife. Many educated individuals look at the simple faith exhibited by persons in impoverished countries and feel sorry for lives lived with such superstition. But the simple journey of some impoverished persons may lead to an adequate understanding and acceptance of the plan that many intellectuals fail to examine. Ironically, in the afterlife it will be many of those simple people who will feel incredibly sorry for the intellectuals who chose to ignore the evidence and who failed to question reality with the necessary diligence.

In 2Peter 3: 8 - 9 Peter states, *"With the Lord a day is like a thousand years and a thousand years like a day.* [Again an indication of how God exists outside of our dimension of time] *The Lord is not slow in keeping his promise, as some understand slowness. He is patient with you, not wanting anyone to perish, but everyone to come to repentance."*

Putting guilt and fear in perspective:

We need to fear God if we have not made a commitment to him to accept His plan. Once we have made such a commitment, the grace paid for by God himself covers us. We should not get mired in guilt if our hearts are moving in a loving, committed direction. John, the disciple of Jesus, clarifies this point in I John 4: 18, 19:

"There is no fear in love. But perfect love drives out fear, because fear has to do with punishment. The one who fears is not made perfect in love. We love because he first loved us."
Therefore, with the fear in life, he gets our attention; and then his perfect prophesied plan gives us the grace and assurance that fear now has a much lesser role in our lives moving

forward. This brutal life and God's design of it and awareness of it should help us to put guilt in its proper place as well. Guilt is debilitating. It prevents us from walking in better understanding and in love with God. It turns the focus too much on ourselves as opposed to God, his provisions, and his desire to work through us for the sake of others in the time we have left. Unfortunately, in many religious communities and churches, guilt is heaped upon people in unhealthy ways. Guilt is one of the first things people turn to when they do not comprehend God's reality. Many walk away from church because of guilt. Again, guilt puts too much emphasis on us as individuals. We should keep our eyes on the love of God and be led by love rather than by fear or by guilt. Remember the words of John: *"fear has to do with punishment."* Obviously, there is a balance to be found here. Our conscience and the leading of the Holy Spirit can convict us when we are in error – which can happen quite frequently. Our response to such nudging can take us immediately back to a healthy, loving relationship with God. To obey [these convictions] is better than sacrifice. The way to progress is typically via a committed, loving desire to please God and renew our thinking to be more effective and to gain the wisdom that assures our minds and hearts that we are moving in the right direction relationally and practically. Guilt, promoted by many religious teachings, has sent people into mental wards and led others to take their own life.

For those who feel stuck in a pattern of guilt, here is a strange but helpful perspective. God does not sin. But he purposefully created the beings that commit the most horrible atrocities. Therefore, why would one be consumed by guilt when God created people he knew would commit the most vile acts? It is quite a point to ponder. And it is a perspective that makes sense. Otherwise, God is not in control. Thus, God is somewhat on the hook here. This life is quite messy. At times it is a hell on Earth. Don't try to hide your sins from God. Instead, take them to Him. He knows exactly the problems commonly faced by people. You have not freaked God out. Don't listen to well-intentioned Christians who say things like God and the Holy Spirit are so gentle that they cannot handle sin and must look away. Those are the kinds of ideas that build with other ignorant ideas to form a tidal wave of guilt that pummels people today. When you confess your sin, and through a spiritual epiphany commit yourself to God through the act of Jesus, God forgives your sin. You become a new creation moved by love and compassion and no longer are to be ruled by fear and guilt. Be moved to tears that God forgives, but do not stay in the mire of guilt. It does no one any good; and it is an unbalanced view of God's universe and his love for each of us. Get over the guilt and "get over yourself". Enjoy the new you that lives in the wisdom of reality. Through Jesus, you can be "born again." It does not make you perfect in this life. It makes you forgiven. It gives you mercy. It replaces fear with love. We should renew our minds to this reality daily, because we have work to do. It starts with us allowing God to get our thinking in line with reality. Do not follow a teaching that encourages you to ignore reality. Your accountability and ability to live with joy and to positively affect others [which becomes, if it wasn't before, your greatest mission] is an accountability to reality rather than emotion.

In bringing this chapter of heavy ideas to a close, an extremely important point needs to be made. The point is this: have some fun. We are to seek a balance of recreation in our lives. When it is time to take a break from work and serving others, take a break. Have a blast. Enjoy your friends and family. Yes, we need to build our spiritual understanding, serve when we are supposed to serve, support ministries and charities, and be there for others. But we are also to take time regularly to relax, enjoy, treat ourselves, our families, and our friends. The sky is more blue, and food tastes better when we move through life in a purposeful manner with God. Tears are more powerful, and laughter is more fulfilling and enjoyable when we work hard and play hard with balance. Don't let fear or the need to prove something or a strange sense of inferiority or obligation incline you to over-commit. People with high empathy that are easily made aware of regrets in their lives are susceptible to over-doing it and to denying themselves too often and too much. Reach out to others for perspective if this describes you. You are a more effective and powerful you when you do what you should and not do more than you should. And you are more powerful and effective when you can recreate and take time to indulge yourself occasionally. Celebration, good food and relationships are a prescribed and important part of your life. If that is missing, lean on others you trust to move you to the proper balance. Just because the afterlife is the ultimate goal and reward, does not mean that the beautiful things God created here are not to be enjoyed.

Balance is important. We should care for ourselves and care about others. Giving and living with purpose are keys to a happy, healthy life.

Chapter 8

200,000 Years of Humans

Humans differ from hominins, and a localized flood is possible both scripturally and scientifically while a global flood is not.

There are things we know and things we don't know regarding the spread of humans across the planet over the past 200,000 years. While archaeologists continue to uncover clues, a fanciful story is prevalent in churches. It is a story of a global flood involving Noah. Such a story is neither scriptural nor scientific. The account in the Tanakh (scripture) is of a localized flood – an account that can easily line up with history and science. The confusion surrounding the flood is similar to the age-of-the-Earth confusion. The Genesis wording of creation appears to be describing the big bang and an accurate scientific progression of the Earth and many of its life forms. Just as some well-intentioned Christians have turned the age of the universe into a tale of less than 10,000 years, many devoted Christians share a tale of a global flood and two of every animal being rescued aboard an ark.

When intently reading the account of the flood in Genesis, one should quickly notice in Chapter 7 verse 2 and 3 that God did not instruct Noah to take two of every animal. The ark was big, but not that big. Noah was instructed to take seven pairs of certain animals, two pairs of others and seven pairs of birds – and by implication, none of most animals. The flood was not global. He did not need to save all the world's species. Life for animals in other locations was likely minimally affected. Noah and his family would be starting from scratch and would require certain animals.

Applying literary and logical integrity to the flood account in the book of Genesis makes it easy to determine that a localized flood is being described. In the poetic Psalm 104, the promise is made by God that, once the waters gave way to the emerging land mass on Earth billions of years ago, *"never again will they [the waters] cover the earth."* (Psalm.104.9) Thus, a global flood in Noah's day would contradict this promise made elsewhere in scripture. In other words, the flood was not global. If it were it would be a scriptural contradiction. Additionally, contextual clues in the Genesis account help to determine that the flood was not global. The phrase that God will *"wipe from the face of the earth every living creature I made,"* (Gen. 7.4) is later clarified in Genesis 8:9. *"But the dove could find no place to set its feet because there was water over all the surface of the earth."* (Gen. 8.9) At that point, the hills were already exposed. Therefore, the surface of the earth could not be the entire planet. Thus, the passage interprets itself. The face of the earth and surface of the earth were only a region of the Earth. Geology and archaeology would not contradict a massive flood on the Mesopotamian Plain sometime in the past 80,000 years.[1] Determining when such a flood occurred is nearly impossible. A scant amount of evidence exists for such a flood in the Mesopotamian Region, because after thousands of years, such geological evidence is difficult to detect.

There were hominins [human-like primates] going back millions of years. In 2015, Prof Brian Villmoare of UNLV reported that a recently discovered 2.8 million year-old specimen of a hominin in Ethiopia was closely related to Lucy from 3.2 million years ago.[2] The BBC reported it as the "first human" discovered in Ethiopia. However, this is a misnomer. There are many species of the genus homo, but humans differ in many significant ways from Lucy and hominins. Most scientists still consider that we modern day humans emerged within the last 200,000 years. Evidence suggests a large group of humans were killed in 70,000 B.C. in Sumatra, Indonesia when a colossal volcano erupted at Toba.[3] And one source claims evidence the human population on Earth dropped to as low as 40 pairs of males and females around this time.[4] Might the confidence interval be off just a bit, allowing for the notion that there were only 4 pairs left: Noah, his three sons, and their wives? Or did God only need to start over with his chosen line of people at that point in history? It is encouraging that today we can determine a bottleneck of humans [a very few humans left on earth] at about the time of a key localized flood.

The account of the Tower of Babel in the Tanakh is also not what many have been taught. First, it reads more like an allegory or legend. It is strangely nestled in between a much more academic flow of thought regarding the descendants of Noah. (Gen.11.1 – 9) The account of the flood incorporates very specific names and details. No specific persons are mentioned in the Tower of Babel. Instead, there is a poetic discussion among the godhead of how they must create different races and languages and scatter the people. The passage ends; and then the text returns to the flow of thought regarding the descendants of Noah. Verse 9: *"From there the Lord scattered them over the face of the whole earth."* The tower being proposed was simply a large structure that would cause the people to *"make a name for ourselves and not be scattered over the face of the whole earth."* (Gen.11.4) They were in Mesopotamia, where large stones were scarce. They had to make bricks from clay.[5] Other ziggurat towers had been built. These were not huge edifices, either.[6] The practical take away from the passage is that God instructed humans to spread around the globe. He knew that sea levels would be low for 30,000 to 70,000 more years, allowing eventually for groups to reach the other continents before the land bridges were covered by water as the ice age ended.

Whatever happened preceding and following a localized flood, humans eventually migrated. Many scientists point to DNA and the notion that indigenous North American peoples migrated from Asia. Many likely settled on the Bering Land Bridge for as many as 15,000 years until rising water levels forced them to continue to North America.[7] Thus, they preceded Europeans in the Americas by thousands of years. As the planet continued to warm from the last Ice Age, the land masses connecting the continents were again covered by sea water. Prior to being covered by sea water, the continents were connected by these land bridges. Thus, the earth was again *divided* once the land bridges were covered, and the continents were once again separated by water.   In the genealogy from Noah to Abraham, there is mention of one ancestor that allows a calibration of time. Genesis 10:25 mentions Peleg and that *"in his time the earth was divided;"* (also 1Chron.1.19) As previously mentioned, generations are skipped for whatever reasons in these accounts. The water levels covered land bridges and once again *divided* the continents a little more than 10,000 years ago. Peleg must have been alive at that point in time.

Genesis 6: 3 declares that, after the flood, humans' *"days would be a hundred and twenty years."* We know today that this declaration is biologically accurate. However, prior to the flood,

The Tanakh and other ancient sources claim that some humans lived hundreds of years. It is nice to stick to what is known and verifiable – such as the Fab Five and the clear plan involving Jesus, but it would be appropriate to have some *what-ifs* regarding this idea of humans living to be hundreds of years old. Here are some possible simple answers to the long-life claims. They range from "Aw, heck no. This portion of scripture is wrong and is not inspired."

To "This, too, is merely allegorical."

To "Maybe they should have put someone else in charge of producing the annual calendars."

But let's consider how claims of long life might be scientifically possible. Perhaps, up until the flood, humans had higher concentrations of enzymes such as telomerase.[8] Telomerase is vital in the regeneration of body tissue. Avoiding meat and not being subjected to as many harmful rays as we are today, could prolong life. Earth has received a significant amount of harmful cosmic radiation from the Vella supernova in nearby space since some time in the past 40,000 years.[9] The timing of this supernova with a flood and an adjustment by God on certain enzymes could coincide with the dramatic change in average life length.

Long life could explain the question some ask about people living in Nod east of Eden. Genesis Chapter 4 describes that after Cain killed Abel, he moved to Nod, had a wife, and had children. Cain was a son of Adam. If Adam and Eve and their children lived for hundreds of years, they would have scores of children, who would each have many children as well. There would be thousands of people in the area. Some could have lived in Nod – hence, the ability for Cain to live there. The first humans were fine-tuned animals living in more ideal conditions and perhaps containing different levels of enzymes. God could have intervened at the flood and at the time when people groups needed to migrate to fill the Earth in time before sea levels retracted. He could reduce the levels of key enzymes and introduce melanocyte cells into the skin to produce varying levels of melanin – creating different races.

The point to be made is this: there are still many things we cannot know for sure. No one should have a problem scientifically with a localized flood in the Mesopotamian Region of the world at some time in the past 80,000 years. No one should be afraid of the evolutionary process which requires intervention by God at key points in the past four billion years. And no one needs to think that scripture and science deviate because someone had to fill in unknown gaps with his or her bias. Let archaeologists and other scientists continue to discover. Meanwhile, the plan provided by God to usher humans into an afterlife is well established in both scripture, history, and science. We can recognize who Jesus is and what this God of incredible life forms and universe has done for us.

Chapter 9

Signs and Prophecies

The information contained in signs and prophecies from the Tanakh point conclusively to a Messiah who would eventually enter Jerusalem on the mission to die for mankind.

Over the years, many atheistic intellectuals have dismissed prophecy. Some make the argument that the Bible was written after the death of Jesus, and of course there would be prophecies and fulfillment written in such a book. The beauty of the way God established all this is what is gleaned when we listen to the Apostle Paul tell us that scripture is the Jewish Tanakh and not the entire Bible. The Tanakh was already in existence at the time of Christ. Thus, these prophecies and signs were all written centuries before Jesus arrived as a man. They are contained in the holy book of the well-established Jewish faith. A priest with access to the temple at the time of the crucifixion, could go read the scrolls describing the events taking place. It is why a spiritual ideology holding accountability to history and science concludes that Jesus is the Messiah and that his death and resurrection pave the way for us. It is important to mention that the birth of Christianity is actually the fulfillment of the Jewish religion.

Many signs and prophecies have been discussed already in this book. In 2000 BC Abraham broke bread and had wine with Melchizedek prior to the establishment of the covenant God made with humanity. Abraham gave a tithe to this person of the godhead – Jesus. The evening before Jesus was arrested and the events were put in motion for him to fulfill the sacrifice required in the covenant, he broke bread and had wine with his disciples.

When Abraham was walking his son up to the altar for the sacrifice, his son asked where the sacrifice lamb was. Abraham replied, *"God himself will provide the lamb."* On that occasion, God provided a ram whose horns were caught in a thicket. This was a sign of Jesus who would have his head in a crown of thorns at his crucifixion. And it was God himself, Jesus, who played the ultimate role as the sacrifice lamb.

In 1500 BC Moses was given instructions regarding the requirements of Passover. The blood from a perfect Passover lamb was to be applied to the tops and sides of the wooden doorways in the homes of Israelites captive in Egypt to avoid the death of their first-born sons. This act closely resembled the blood of Jesus on the wooden cross 1500 years later. God shared through Moses that in future Passover celebrations the lamb should be selected on the 10th day and sacrificed on the 14th day in preparation for the special Sabbath on the 15th day. This signified Jesus eventually coming into Jerusalem on the 10th day of Passover, being sacrificed on the 14th day of Passover, and being taken down off the cross before dark, because the 15th day was the special Sabbath. Moses and – 500 years later – David both declared that the Messiah's legs would not be broken. When Jesus was crucified with two thieves, all three men needed to be taken off their crosses before sundown due to the special Sabbath. In order to accomplish this, the legs were broken to prevent a person from pushing up to get air. But Jesus was already dead, so his legs were not broken – fulfilling the prophecies.

The gospel of John records in Chapter 19 the scene at the crucifixion of Jesus. Matthew 27, Mark 15, and Luke 23 also record events. Jesus cried out, *"My God, my God. Why have you forsaken me?"* Soldiers had circled him and beat him to near death prior to nailing his wrists and ankles to the cross. They cast lots for his clothing. Had a priest walked from where Jesus was being crucified into the temple to the scrolls comprising the Tanakh, the priest could have read all of what he just saw in the 22nd Psalm of David written 1000 years earlier - beginning with the words *"My God, My God. Why have you forsaken me?"* That same priest could then go to the scroll containing Isaiah's writings from 700 BC and read in Chapter 53 another detailed description of the crucifixion that was taking place. He would read about the Messiah being rejected by men, being crucified with criminals, and being buried in a rich man's tomb – which would occur in a few more hours. He would read of a Messiah who had to be pierced. Not only did Isaiah prophesy his being pierced, but so did Zechariah. This was done at the crucifixion to the lung to confirm that blood had mixed with water to fill the lungs and kill the person on the cross. Isaiah's prophecy declared that the Messiah would go through this willingly and not defend himself and that he would rise from the dead and see the light of life again. *"He would bear the sins of many and pour out his life to death."*

The priest could then go to the book of Deuteronomy [Chapter 21 and verses 22 and 23] and read in the law that God would curse anyone who is hung on a pole – on a tree – such as Jesus. This is how God could justify cursing Jesus with all the sins of mankind. This priest could then make the connection to the Passover lamb and the blood on the wood in the pattern of a cross that would prevent the second death of men and women.

The priest could read about cities of refuge [Numbers 35: 6 – 34] – how persons accused of murder could run to a city of refuge, hold on to the horn of the altar of sacrifice, and plead for innocence. If it was determined the death was an accident, the individual had to remain in the city for the event that would wipe away any guilt and allow the individual to return home. The event to free him was the death of the High Priest. Combining this with David's words that the Messiah would be a high priest in the order of Melchizedek, the priest could make the connection to the death of the high priest. This would be confirmed in Psalm 118: 27, discussing he who would come in the name of the Lord, and the invitation to those in the Messiah's time: *With [tree] boughs in hand, join in the festal procession up to the **horns of the altar**."* The priest at the time of the crucifixion could then recall the precession that occurred four days earlier – on the 10th day of Passover – ending now in the sacrifice of the ultimate high priest. He could recall how Jesus entered Jerusalem on a donkey, the sign he was king according to the prophecy in Zechariah 9:9.

These prophecies and signs have already been discussed in this book. But there are many more prophecies and signs contained in the Tanakh. Some of them cannot be confirmed with certainty but are generally held to be true. This includes the incredible unlikelihood that the Messiah would be born in the small town of Bethlehem. It is difficult to substantiate the claim in the Gospel of Luke [Luke 2: 1] that Caesar Augustus declared a census and that residents had to make a pilgrimage to the town associated with their family line. This required Joseph to leave Nazareth and go to the little town of Bethlehem, where

Jesus was reportedly born. The prophecy of the Messiah being born in Bethlehem comes from Micah 5: 2.

"But you, Bethlehem, Ephrathah, though you are small among the clans of Judah, out of you will come for me one who will be ruler over Israel, whose origins are from old, from ancient times."

But while historians have difficulty substantiating the census decreed, the logical conclusion is that it occurred. The physician, Luke, included it in his gospel. Had it not been true as people of the time would know, it would discredit the rest of what he had to say. Thus, logically it is a near certainty that Caesar Augustus declared such a census.

The birth of Jesus claimed in Bethlehem has additional significance to the prophecy. This was a town with many shepherds, who were among the lowest social class in Judaism due to their work with animals and having contact with sheep feces. It implies that after 13.7 billion years of preparation, God chose to be welcomed by the most lowly people on the planet. This confirms the notion that God loves the humble and resists the proud. Additionally, many of the lambs used annually in Jerusalem for the Passover were likely raised in Bethlehem.

Daniel prophesied the time of the Messiah written 500 years before Jesus arrived. In Chapter 9 of his book he intimates – using weeks as seven-year periods – how many years it would be from the decree to rebuild Jerusalem until the time of the Messiah. Historians using various estimates for the times of those decrees and using estimates of the precise year of Jesus death can confirm that the death of Jesus occurred within the window of legitimate times. While there is no way to confirm the exact fulfillment, we know it is at least extremely close and why people around that time were looking for the Messiah.

One of the most descriptive and jaw-dropping prophecies involves the betrayal of Jesus by a close friend. Matthew Chapter 27 records the actions of Judas. Judas agreed to betray Jesus for 30 pieces of silver. But after Jesus was captured, Judas became distraught and went to the temple to return the money to the high priest. The high priest would not accept it, so Judas threw the 30 pieces of silver back to the high priest in the temple. He eventually hung himself in a potter's field. The priests could not keep the silver, because it was blood money. Thus, they decided to pay the potter for his field, which many considered cursed since the betrayer of Jesus hung himself there. These details of how the Lord was betrayed were prophesied in the book of Zechariah, Chapter 11: 12 - 13:
"I told them, 'If you think it best, give me my pay; but if not keep it.' So they paid me 30 pieces of silver. And the Lord said to me, "Throw it to the potter" - the handsome price at which they priced me! So I took the 30 pieces of silver and threw them into the house of the Lord to the Potter." Of note, is that the JPS translation of the Tanakh to English uses the term *accountant* in place of *potter*. However, the Hebrew word is *yatsar* and is translated *potter* and not *accountant* in six other Tanakh passages, including Isaiah 41:25, *"he treads on rulers as if they were mortar, as if he were the potter treading the clay."* It is understandable that Jewish translators would question how a potter would be in the temple – hence, the translation to an accountant receiving the 30 pieces of silver. This is why it is one of the more impressive prophecies. How were the coins – the ransom paid for betrayal of the Lord – going to be thrown into the temple to pay some potter?
Another sign involved Moses after he led the captive Israelites

out of Egypt. This large group of people were in the desert, needing water. Many were angry at Moses and his brother Aaron. Exodus 17: 4 records Moses crying out to the Lord, *"What am I supposed to do with these people? They are almost ready to stone me?"* Exodus 17 continues, *"The Lord answered Moses, 'Walk on ahead of the people. Take with you some of the elders of Israel and take in your hand the staff with which you struck the Nile, and go. <u>I will stand there before you by the rock at Horeb. Strike the rock, and water will come out of it for the people to drink</u>.' So Moses did this in the site of the elders of Israel."* [Ex 17: 5 – 6]

A year later the group was back at the same location. The people again grew angry and were without water. Again, Moses and Aaron were frustrated and turned to the Lord. Numbers 20: 6 – 13: *"Moses and Aaron went from the assembly to the entrance to the Tent of Meeting and fell facedown, and the glory of the Lord appeared to them. The Lord said to Moses, 'Take the staff, and you and your brother Aaron gather the assembly together. <u>Speak to that rock</u> before their eyes and it will pour out with water. You will bring water out of the rock for the community so they and their livestock can drink.'*
So Moses took the staff from the Lord's presence, just as he commanded them. He and Aaron gathered the assembly together in front of the rock and Moses said to them, 'Listen, you rebels, must we bring you water out of this rock?' Then Moses raised his arm and <u>struck the rock twice</u> with his staff. Water gushed out, and the community and their livestock drank. But the Lord said to Moses and Aaron, 'Because you did not trust in me enough to honor me as holy in the site of the Israelites, you will not bring this community into the land I give them.'"

This seems harsh on God's part. Moses risked and accomplished much. Why did him losing his temper prohibit him from entering the promised land with the congregation? One issue may have been Moses stating, *"Must we* – implying Moses and Aaron – *bring you water out of this rock?"* This may suggest Moses took credit for the water. But there is a far more important issue. Like Abraham meeting Melchizedek and breaking bread before the establishment of the covenant, the interaction with the rock was to be another sign of the Messiah, Jesus. The Messiah would be struck down once for mankind. Hence, initially God instructed Moses to strike the rock once. With the second interaction at the rock, God commanded Moses to speak to the rock and water would come. This is a foreshadow of how the sacrifice and resurrection of Jesus would allow men and women to pray to God, and God would grant his Holy Spirit to live within them. But Moses failed to follow through with this order from God, destroying the poignant symbolism. God had to make a big deal out of this to preserve this important sign. Paul stated 1500 years later that the rock accompanying the Israelites in the dessert was Christ Jesus. [1 Cor 10:4] Notice in the first instance of Moses and the rock recorded in Exodus that the Lord stated, *"I will stand before you by the rock at Horeb."* The plan of God was precise, with multiple signs, actions, and prophecies pointing to it. From a mathematical perspective, there is no way to expect these fulfilled prophesies are mere random coincidences – not when they are found in the one most commonly accepted scripture on Earth and when they contain so many vivid details of the Messiah, his purpose, and what he would endure. It suggests the simplest and most logical explanation for the presence of human beings on Earth.

Knowing the existence of the Tanakh and its incredible detail of events that would happen centuries after it was written, we know that humans have been contacted by a being who exists beyond our space-time dimension. With this condition established and applying conditional probability to the explanation for the existence of humans on Earth today, the strongest scientific conclusion is that this triune God created the universe and we humans. The details of how He used evolution and natural progression along with moments in time where he intervened with mass extinctions and mass repopulations cannot be precisely known.

Knowing how rare it is to find a planet with Earth's physical characteristics and beneficial properties and knowing how close Earth came at various times to becoming sterilized, it is not rational to conclude that human life on earth is a product of random chance. Another important point should be made. Usually more than once per year, NASA locates a new planet in a habitable zone in nearby space that *might be habitable.* This does not mean it has many of Earth's characteristics. It merely implies it is located in the habitable zone of its star(s). It may be small and rocky with water, but it does not imply there is life, and it does not imply that animals could be transported and survive there.

Recall that Naturalism is what is taught as science in American public schools. But Naturalism is a philosophy and not pure science. Pure science makes no assumptions. However, Naturalism assumes that no influence can exist outside our space-time dimension. The argument against naturalism is strong and damning. There is currently no explanation for the origination of self-replicating material on earth. Naturalism cannot offer explanations for mass, sudden repopulations that followed significant extinction events and seemed to anticipate future environmental conditions. Some ideas still promoted in classrooms have been discredited for years. Such is our existence in a time when so much information is available. We as humans – regardless of intelligence levels – feel safe when we belong to a large group of like-thinking people. It is yet another example of the human-nature dynamic created by God. Can an individual get past the stumbling blocks of human nature? Coming to grips with the reality of this life requires humility and integrity. And it requires one more component – a spiritual component. People can read a book like this, read of the fulfillment of prophecies, see the frailty of Naturalism, and be shown the plan and the steps to acquire this gift of life-after-death. Yet, many will not process it all. We are complex beings with emotional, intellectual, and spiritual components. The humility and integrity must occur on all three levels to have the eyes of our understanding open to such a brilliant plan and such a brilliant and loving God. Much like financial and business situations can be viewed from the standpoint of a game, so can life itself with its true purpose. Solving life's meaning – and thus winning the game – requires effort and virtue. And often the revelations and understandings require pain and suffering. We are good and we are bad. We are wise and we are ignorant. We are powerful and we are susceptible. Just like the physical universe and physical make-up of life are beautifully complex and precise, so is our very existence from birth until death. The interactions and experiences in our lives have great importance in our current existence as well as discovering, accepting, and acquiring the life yet to come.

Chapter 10

Ideological Suggestions

This final chapter includes sound doctrinal ideas in primarily chronological order, how to accept and follow God, and how to live for those who would like such direction. Transforming our lives for the better is a process that takes years. Key in the process is the attempt to be life-long learners – continuing to question, listen, reason, and insist on accountability. The journey and the process are as important as the wisdom gained. Many value a perspective that lines up with reality and is based upon love and compassion and respect. Living organisms involve incredible precision. The universe involves incredible precision. This book is a call to spiritual precision. **Understanding Scripture and the New Testament becomes much easier once an individual can see the greater context of life and the message and plan designed by God.**

In addition to precision and logic, God is also the God of relationship and compassion. He is a God who breaks rules when it is the right thing to do. In fact, He established a set of rules that can entice some who lack integrity to think the letter of the law is superior to the spirit of the law. But it is God, his love, his grace, and his promise that are superior to the letter of the law.

The message is to overcome one's *ego, bias and self-centeredness* [EBS: the *flesh*] in order to recognize the existence of an afterlife, the provision of the afterlife, and the ultimate purpose of this life. This life with its beauty, brutality, fascination and mystery has the utmost purpose. The Tanakh is clearly inspired from an intelligent source beyond our space-time dimension. Jesus [Melchizedek], part of the godhead, is exactly the God we would want him to be. He was welcomed into this world by the lowest class of people – shepherds. He worked with his hands for thirty years. He displayed to the world who he was via his compassion toward the sick and the outcast and the misunderstood, and with his respect for women, children, and the elderly. He confronted religiosity and conventional wisdom with a godly, practical one. Unfortunately – yet by design – followers over the years have altered this wisdom. We are all human and subject to making mistakes of varying degrees. The plan contained in the Tanakh, lived out by Jesus and outlined by Paul, is clear. We do not get to earn our way into eternity. We simply appreciate our way into eternity by recognizing, accepting, and living the plan.

Doctrine in this book leans heavily on the Apostle Paul and on logic and a more mathematical approach to rule-out impossible doctrine and to support more accurate interpretations. The teachings of Paul and Jesus are very similar and paralleled. In case it has not been clear, the Bible includes the Tanakh [Scriptures, Old Testament], and the New Testament record of the life of Jesus [the Messiah] and of important instructions and interpretations for life. Like the Tanakh, there are some prophecies in the New Testament which may not make much sense currently but will be vital at some point in the future for people who will live through the most difficult time in the history of mankind. Recall, also, that Islamic faith gives special emphasis to the first five books of the Tanakh (Old Testament) and to the Psalms (David's writings in the Tanakh). It is perhaps why God features so many foundational things and signs and prophecies in these six books.

Quotes here tend to be from the New International Version of the Bible. No translation is a perfect translation from the original Hebrew of the Tanakh or the Greek of the New Testament. In fact, at various times, well-meaning translators injected their bias into their translation. It is important to attempt to hold accountability to the original Hebrew and Greek and the likely interpretation. Additionally, The Tanakh [Christian Old Testament] is spiritually discerned and can be veiled – requiring diligence and humility. (2 Corinthians 4. 1 – 8). We pray for wisdom and understanding. It is the Holy Spirit who opens our eyes spiritually. But be careful. This is not a license to believe anything we assume is from God. If our ideas and interpretations cannot mesh with scripture and the teachings of Paul and Jesus, then why would we assume that they are insights from God?

2Cor: 4: 1 - 8; John 7: 37 – 39; 1 Corinthians 1: 18 – 2: 16

What follows are several key doctrinal ideas. This is not an exhaustive list – meaning there is more involved. It follows a primarily chronological order.

Triune God: In the Tanakh, the Jewish holy book for a Jewish religion that believes in one God, it is impressive to see the quote in Genesis [1:26]] *"Let us make man in our image."* This implies one God with more than one personality. In John [John 17: 1 – 5], Jesus is praying to the Father God and discussing how they made plans before the beginning of the world.

Gen: 1:26; John 17: 1 - 5; John 1: 1,2; John 11:41,42

Space-time dimensions: In 2Tim 1:9,10, Paul tells us that, *"The grace given to us in Christ Jesus was given to us before time began."*   Genesis 1:1: *"In the beginning, God created the heavens and the earth."*   Many scientists as recently as the beginning of the 20th century still assumed that the universe was infinite and static.   Many religious ideologies believed this as well.   But the Tanakh accurately acknowledged that our space-time universe began at some point in the past.   And as the author in Hebrews [11: 1 - 3] accurately reported, the universe was made from that which is unseen.   Subatomic particles cannot be seen with the naked eye or with powerful microscopes, but God created with material not seen.   And God must exist in at least two more dimensions of space and two more dimensions of time compared to the space-time dimension of our universe to be in control of both our physical realm and a spiritual realm.

The prayer Jesus made to God when Jesus was on earth is also an indication of this.   John 17: 4, 5: *"I have brought you glory on earth by completing the work you gave me to do.   And now, Father, glorify me in your presence with the glory I had with you before the world began."*

    Gen: 1:1; Hebrews 11:1 - 3; John 17: 4-5

The Genesis account of creation, space, and Earth describes six long periods of time – one of which is more than one billion years long.   The Hebrew word for day is yom [or yowm] and can mean period of time.   The passage interprets itself with various contextual clues, including how there was *"evening and morning"* of each day.   But on the seventh day [Chapter 2], there is only evening and no morning recorded.   Hence, we are still in the 7th day of creation – 7th long period of time.   God is still very active but has rested from creating new things.

Instead, species adapt and change. They evolve over time.
Gen: 1:1 – 2:3; Psalm 90: 1 – 4

Unlike other animals, humans are part spirit. This is what was meant when God said, *"Let us make man in our image."* We have a spiritual component in addition to a physical body. When we are *"born-again"* spiritually, it implies we are born into a heavenly realm that exists on Earth now and will continue for eternity in the presence of God.
Genesis 1: 26; Luke 17: 22 – 37

God exists in a realm that includes and is outside of our space-time dimension. This is why God described himself to Moses as *"I AM"*. And this is why the author of Hebrews declares Jesus Christ as *"the same yesterday, today and forever."* Nothing gets past God. He cannot be fooled by beings he created as a result of this. This implies more of the doctrines that follow, and this is how **God is in control while humans have a free will.**

The death that came to mankind as a result of Adam's sin was a spiritual death. Many animals died prior to human presence. It provides the fossil fuels for today's civilizations. Humans were not created perfect. They were created with good and bad tendencies that make it possible – for those who fight human tendency and seek with humility and integrity – to determine that we are flawed and yet God loves us so much that he has prepared an afterlife and paid for the passage. Death physically was inevitable in this finite universe that would eventually burn out and die if left to itself. Human flesh and the universe were not created for eternity. The human spirit was.
Exodus 3: 14; John 8: 58; Hebrews 13.8

God created a spiritual realm of beings before He created animal life. Jesus shared what he saw long before he came to Earth as a man. *"I saw Satan fall like lightning from heaven."* [Luke 10:18] We are told in John 16: 11 and John 12: 31 that Satan is the prince of this world and that we wrestle with principalities and powers in the heavenly realm [Ephesians 6: 10 & 11]. In scripture, there is an account of an angel being sent to answer a prayer from Daniel. The angel was intercepted by the prince of Persia – a spiritual being: an evil spiritual being. Finally, the archangel Michael came and freed the angel to get to Daniel. **Conclusion: God created beings he knew would turn and do evil.** This was part of his purpose for this life and the need for us to exist in a universe with pain, suffering and evil.

Hell is a real destination prepared for the Devil and his angels, and humans old enough to make a decision may ultimately decide to join Satan in this place of torment. God is not willing that any humans end up there, but many will. It is part of the brutal necessity of this short life in preparing an eternal life of optimal interaction. People in the afterlife will have grateful hearts and be given a new body not prone to simple negative human traits. The combination of beautifully committed and transformed souls and an existence in a more dynamic space-time existence makes heaven more incredible than we can fathom. The precise details of hell are not fully known. Jesus tells us to fear it and to respect [fear] God who will send us there if we ignore His provision.
2 Luke 12: 4, 5; Matthew 25: 31 – 46; John 3: 1 – 21

The universe is billions of years old as is Earth. Fossil evidence clearly establishes this fact. Death of animals and plants millions of years before humans arrived prepared a planet for humans that was rich in fossil fuels, allowing us to investigate and learn and enjoy a better quality of life. This book explains much of this in detail.

Recordings of past individuals and their lives and struggles help us today to recognize the good and bad in us. The majority of great men and women recorded in the Tanakh made major mistakes. Yet, God used them and created signs from their lives to help seekers realize the validity and inspiration of his plan to come to Earth and be sacrificed (killed) for our sins. Abraham's impatience produced two blessed people groups that have warred against each other throughout the centuries. Moses killed a man. David committed adultery and then devised a plan that would cause the incredibly devoted husband of this woman to be killed in battle. Saul [later Paul] hunted early Christians – imprisoning some and killing others. None of us want to do bad. We should be careful not to assume we are right. When God's followers get reckless, it can lead to enormous consequences. Nothing should be taken for granted. It is healthy to admit humanness and to focus on doing good whenever possible. The wonderful recognition of God's grace, love for us, and desire to use us to better the lives of others is to recognize that we cannot lose our salvation when our heart is in the right place in making a commitment to him and attempting to persist.
Ephesians Chapter 2

The recording in the Old Testament of a massive flood gets many different reactions from people. Most assume it is too far-fetched and must be a symbolic tale or evidence that

scripture is ridiculous. Others believe it is a literal depiction, that it covered the entire planet to the tops of the highest mountains; and some believe it happened less than 6000 years ago. But as eluded to earlier in this book, the Genesis account of a flood interprets itself. It was not global. It was regional. This is in line with the account in Genesis 7 & 8. It is also in synch with Psalm 104, which suggests that – after the initial emergence of land on Earth billions of years earlier – never again would the seas cover the tops of all mountains. If an actual flood covered the entire planet to the tops of the mountains in Noah's day, the Old Testament Tanakh would contradict itself. God did not instruct Noah to take every kind of animal aboard. It was a limited list of animals. After all, a flood that was not global would not disturb the majority of animal life on Earth at the time. According to scripture, men had become incredibly evil. God caused a flood of what would be the Mesopotamian Valley perhaps 40,000 to 80,000 years ago. Given the dates of the last ice age, this is a possibility scientifically. Literary integrity implies that God allowed a flood to wipe out mankind from one particular region of the globe. The Biblical text cannot allow us to conclude without question that no other region of the world had humans dwelling in it. There are some questions without answers currently.

Psalm 104: 1 – 8; Genesis 7 & 8

Estimating the time at which events like the flood occurred is difficult due to generation-skipping in the Tanakh and in the Christian New Testament. The Hebrew word *abba* can mean ancestor. We see in Matthew Chapter 1 and Luke Chapter 3 that there are generations skipped in the genealogy of Jesus. [see chapter 1] The same situation occurs in Genesis regarding generations from Adam to Abraham. Impressively, we can pinpoint a very significant time in these genealogies from the statement made in Genesis. In Genesis 10:25, one of the descendants was called Peleg. *"In his time the earth was divided."* This lets us know that Peleg must have lived a little more than 10,000 years ago when the land bridges connecting continents were covered up due to rising sea levels as the planet warmed from its latest ice age. Continents were again divided.

Compare genealogies in Matthew Chapter 1 and Luke Chapter 3; Genesis 10:25; 1 Chronicles 1:19

Jesus, also known as the heavenly high priest, Melchizedek, came to Earth and met Abram [later called Abraham]. They broke bread and shared wine, and Abram gave a tithe to Melchizedek. This happened prior to Abraham being willing to sacrifice his son: before the establishment of God's covenant with man. Centuries later, Jesus would again break bread and share wine – this time with his disciples – prior to the fulfillment of the eternal covenant with God through the sacrifice of Jesus.

Genesis Chapter 14; Matthew 26: 26 - 29

God established a covenant with Abraham. God found a man so devoted that the man was willing to sacrifice his own son to God. This faith was credited to Abraham as righteousness, and the covenant was sealed between God and man. The promises and blessings made to Abraham can be ours through faith in God's plan alone – faith in Jesus and what he accomplished. In the recording of the events surrounding Abraham taking his son to the alter to be sacrificed, there are many clues pointing to the Messiah [God, the Son] and what he would suffer and accomplish 2000 years later. Abraham's son asked Abraham where the lamb was for the burnt offering. Abraham replied,

"God, himself, will provide the lamb." This statement had double meaning. First, it implies that God himself will be the sacrifice lamb in the form of Jesus 2000 years later. Second, on that day a ram was found with its horns caught in a thicket. It was sacrificed. It was a foreshadow of Jesus having his head in a crown of thorns at the time of his crucifixion.
Genesis Chapter 22

Jesus completed the covenant by being beaten and hung on a cross
Galatians 3: 13, 14; Romans 10; John 19

God used Moses to deliver his people from slavery in Egypt. The final plague that got the Hebrews released was the first-born son dying during the night. The only way for death to pass over the home was to sacrifice a perfect Passover lamb and put some of its blood on the sides and top of the wooden door frame – a foreshadow of Jesus, the Passover lamb, having his blood placed on the wooden cross to allow those who put faith in God's provision to have the second (eternal) death *pass over* them.
Exodus Chapters 11 & 12

God gave the Law through Moses. [see chapter 5 in this book]
The Law accomplished the following:
It set a standard no human could attain.
Humble individuals could recognize this and investigate how
    to be considered acceptable in God's eyes.
Self-centered, egotistical people would fall into the trap of
    convincing themselves they could live by the Law
    and be acceptable in God's eyes.
Others would think that the law is so ridiculous (which may

seem true) that the notion of a god being behind it
would be nonsense.
It established laws allowing God to sacrifice God the Son
and curse him with the evil done by anyone during a
lifetime. *"Cursed is anyone hung on a tree."*
It provided prophecies and signs that would confirm God's plan,
existence and faithfulness.
> *Deut 21: 22 - 23*
> *The Law is the set of rules and regulations dispersed*
> *throughout the first five books of the Tanakh [Old*
> *Testament]*

**God influenced men from Job prior to 2000 B.C. through some
of the prophets in the centuries preceding the arrival of Jesus
to write the scriptures, the Jewish Tanakh.**
Signs, prophecies and God's control of history would provide
the proof of His existence and His plan and give men principles
from which to live and from which to anticipate the Messiah's
arrival and provision for eternity.
> *2 Peter: 1: 19 – 21*

**The Jewish Tanakh is the "inspired scriptures" to which the
Apostle Paul referred.**
> *2 Timothy 3: 14 – 17; 2 Peter: 1: 19 – 21*

Jesus arrived and fulfilled many signs and prophecies
He is God and human according to many indications. John
describes this well in John Chapter 1
His mother, Mary was a virgin. Jesus was conceived of the Holy
Spirit. This was prophesied in Isaiah 7:4 and recorded in the
Gospels of Matthew and Luke.
His death and resurrection provide the portal for all to the

afterlife
His act allowed the Holy Spirit to dwell in the hearts of believers
and not just in the Holy of Holies of the temple or tabernacle.
He taught his disciples for three years prior to his death. After
his death, the disciples traveled around the known world,
spreading the news of what was done for Jews and for
Gentiles (as was prophesied).
John Chapter 1; Isaiah 46.9; Matt 28.19

**Before the beginning of time, God devised a plan allowing
those who accept the grace of Christ to dwell with Him in a
superior realm referred to as heaven. He influenced Moses to
write the law that *"Anyone hung on a tree [pole] is cursed."***
Paul explains how Jesus was cursed on the cross in this manner
and is our provider in the afterlife.
2Tim 1:9; Deut 21: 22 – 23; Gal 3: 10 - 14

Jesus rose from the dead as prophesized by Isaiah and others.
Isaiah 53: 9 – 12; Psalm 16; I Cor 15: 12 – 20;

**Christianity is the fulfillment and continuation of Judaism. The
Tanakh states that Gentiles along with Jews would reap the
reward of the covenant.**
Psalm 22: 27 – 31; Isaiah 49: 6

**Writings in the New Testament record the fulfillment of
scripture and teachings of Christ and the Christian faith as the
completion of the promise made to man by God.**

**God equips us with everything we need if we allow Him to do
so. Even faith is a gift from Him.**
Ephesians 2: 8; Ephesians 6: 10 – 18

As promised in Scripture, the Holy Spirit now abides within believers as opposed to only dwelling in the tabernacle of God in the days of Moses. He nudges us in the right direction. He gives us wisdom when we need it. He comforts us. He was promised to us by many prophets, and the delivery was made possible by the sacrifice of Jesus.

The Holy Spirit [The Comforter] was promised to the disciples of Jesus and to all future followers. The early disciples were told to pray and wait in Jerusalem until he came. The Book of Acts records His coming on Pentecost. It stated that He came like a rushing wind. [Acts Chapter 2] And when the disciples took to the streets to share what happened, they spoke in different tongues – different languages – so that foreigners in Jerusalem heard the message of Jesus and God's plan in their own native tongue.

There are powerful gifts the Spirit gives to different people to make the Church function more efficiently and effectively in its mission to share the gospel message with as many people as possible. [I Cor 12]

A word of caution: some churches can take the actions of the Holy Spirit to extremes – and some seem to pretty much ignore Him. There are some well-intentioned people and some manipulative people who can create a complicated, emotional set of practices involving the Holy Spirit. Be careful when people say, "God told me this," or the "Spirit is leading me in this direction." Other than searching scripture and the New Testament to see if it supports the claims of such individuals, there is little available in the way of accountability to test for trustworthiness. We should practice humility and state we *believe* God to be leading us.

113

While the Holy Spirit himself is the most powerful force in our everyday lives, teachings surrounding the Holy Spirit can be some of the most destructive in the world today. We should always err on the side of love and accountability to scripture and biblical teaching. Just because people we know may be good and kind in general, does not imply that they have a powerful take on what the Holy Spirit is suggesting for their life or ours. We should respect the spiritual realm and recognize how fallible we are in attempting to interpret signs in our lives and the correct actions to be taken.

The good news is that – in addition to us accepting the gift of eternal life – God provides the Holy Spirit to guide us and take care of us in our daily lives.

> *John 7: 37 – 39; John 14: 15 – 17;*
> *Ezekiel 36: 25 – 27; John 16: 13 – 15; Romans 8: 9 - 16*

Paul – after his remarkable conversion from being the persecutor, Saul of Tarsus – became the intellectual who explained the scriptures, the plan, and key details and perspectives to instruct the billions who would eventually seek Christ.

> *Acts Ch 9; all of Paul's letters [Romans, 1&2 Corinthians, Galatians, etc.]*

Prayer is a powerful way for believers to connect directly to God through the Son. And Jesus prays continually for his believers on Earth. He taught us a model for prayer that includes the components of how we should communicate with God and recognize our relationship with him and dependence on him. See the Lord's Prayer [Matthew 6: 9 – 13]

> *Romans 8: 31 – 35; Hebrews 7: 25; Matthew 6: 5 – 15;*
> *1 Thessalonians 5: 16 - 18*

As God, Jesus knows our human condition well. His teachings, often shared in stories or parables, help us to understand important perspectives on life.
He taught the story of the Good Samaritan (which seems to elude many today). In it, he portrayed how many self-righteous people avoid looking out for others. Yet, a Samaritan – from a group of people looked down upon in the culture of that time and area – was the one who showed a tremendous capacity to care for a total stranger.
Luke: 10: 25 - 37

This life is designed to have tribulation and struggle; and we grow as we persevere through it wisely.
God made no mistakes because he is in the future and the past simultaneously and so cannot be fooled. He made a complex universe with pain, suffering and evil playing an important dynamic in getting the attention of people and in displaying the essence of love as people rise above challenges to be compassionate and dedicated to others.
John 16:33 [in fact, John 16: 17 – 33. He is speaking to his immediate disciples] James 1: 1 – 18
[In fact, all of James Chapter 1 is beneficial]

Baptism is an important display of one surrendering his or her earthly, carnal being and then accepting the gift of rebirth into a life led by the Spirit. We are to be baptized in the name of the Father, Son, and Holy Spirit.
Matthew 3: 13 – 17; Acts 2: 37 – 41; Matt 28: 19

The love of God, exemplified by Christ, and the fruits of the Spirit are now the spiritual law of the day and accomplish what the law given to Moses could not. A person's ability to practice humility and integrity despite our human tendencies plays the greatest role in us searching for the reality of this life, the existence of God, the recognition of our condition, and the plan and the gift paid for by the grace of God. It adds great purpose to this life and is therefore the source of great joy and fulfillment.

Galatians Chapter 3; Galatians 5: 16 - 26

Jesus summarized the law and the prophets – how we are to live – in the simple statement, *"Love the Lord your God with all your heart and with all your soul and with all your mind. This is the first and greatest commandment. And the second [greatest]* **is like it.** *Love your neighbor as yourself."*

Mark 12: 28 - 31

Each person's heart is the determining factor in salvation and in effective service through love rather than obligation.

1 Samuel 16: 7; Luke 17: 20, 21; Matthew 6: 19 –21; Matthew 25: 31 - 46

Confessing Jesus and loving God through one's actions are the display and litmus test for salvation.

Romans 10: 8 – 13

Paul suggests in Romans 2: 12 – 16 that some will enter the kingdom of heaven without hearing the gospel. They will be judged by how they followed their heart. It is important to note that it is still the sprinkling of the blood of Jesus that saves them.

Faith and love (and the good works that come from these motivations) are the desired works as opposed to when people do good things out of obligation or selfish gain. God loves it when – as much as possible – our good works are done for others with few, if any, strings attached.

> *Ephesians Chapter 2*
> ****Paul summarizes much of what we have discussed in this Chapter of Ephesians.*
> *1 Corinthians 13 describes many aspects of love.*

There is a very organized evil spiritual realm. While accounts in the New Testament refer to instances of demon-possession, there is a far greater power at work on the planet. In Paul's letter to the Ephesians, he states we wrestle against principalities and powers. God created these beings to make the trial of this life complete. Subjecting humanity to this beautiful, yet brutal life places humans in sometimes desperate situations. When we get to the next life, we will recognize the benefit of experiences here. Emerging through what can be described as *trial by fire* will produce dynamics in an afterlife of free-will beings not attainable otherwise.

The combination of these spiritual powers working in governments, schools, churches, military, the financial sector, and other groups along with the flaws and inclinations of human nature make the dynamics of life tenuous. Countries can move from relatively healthy cultures to unstable ones in short periods of time. Promotions of thought can be organized by these spiritual forces. Ignoring their existence has consequences.

> *Ephesians 6; Daniel 10*

There will be a day of judgment spiritually in which God will separate the faithful from the non-believers. There will be horrifying agony for those who will be cast to hell with the fallen angels.
Matthew 25: 31 – 46

The afterlife will be a place God is preparing for us. We only arrive there by accepting God's redemption through the sacrifice of Jesus, the Messiah. It is a new heaven and new Earth and a new body in a greater space-time dimension. Every aspect of it will be remarkable – from relationships, to physical dynamics, to fulfillment and being in the presence of God. *"'What no eye has seen, what no ear has heard, and what no mind has conceived' – the things God has prepared for those who love him."* [1 Cor 2: 9]
John 14: 1 – 3; Rev 21: 1

The Earth and this universe will ultimately be destroyed. A new space-time dimension with a new Earth will emerge. When a person dies, he or she is ushered to this different space-time dimension which already exists. It is impossible for any of us to conceive it due to its multiple dimensions of space and time. We as spiritual beings created in God's image will reap the reward of such an incredible life through the faithfulness of Jesus and our acknowledgment of him.
Luke 21: 33; Matthew 5:18; Prophecy in Revelation 21:1

Many signs and prophecies are highlighted in Chapter 9.

PLENTY OF ROOM FOR DISAGREEMENT:
There is much room for disagreement among religious bodies regarding many of the details of life and how to live. Therefore, there are some doctrines not discussed. The encouragement is

to stay true to literary, historical and scientific integrity. Stay true to the spirit of the law of God and the love He promotes. We have an awesome responsibility. It is our responsibility to read our holy book, stay true to established truths and keep an eye on the priorities of life. Our greatest responsibility is to help others recognize who God is and what He can do for everyone. Our inaccuracies can push people away from God. We should study well and struggle well.

How to become a follower of Christ and accept the gift of eternal life and of the Holy Spirit:
God's plan to redeem us all and allow us to join him in the afterlife is extremely precise. It requires laws and signs and prophecies and precise actions. Fortunately, God is loving and practical. He looks on the heart and does not get hung up on formality. When we have the realization of the existence of God and of the death of Jesus being the act of grace that provides eternal life, it is time to act on it and show our gratitude and allegiance to God. If we have not yet had the epiphany but are close, we should pray something along the lines of the following.

Dear God. I am beginning to recognize who you are and what you did for us and how much you love us. I want to have the faith you promise to believe in you and in your Son and in the Holy Spirit. Please overcome my bias and my pride and give me the knowledge and wisdom to understand my human presence in this life and the provision you have made for me regarding eternity. Help me to have the determination to seek. Please help me to read the Bible – which includes the Tanakh, known as the Old Testament, as well as the New Testament – in

119

order to gain the knowledge (and faith) necessary.
From there, re-read this book, and read the passages referenced from the Bible. Always choose to build your foundation on ideas presented and established through discerning the scripture and the Bible. Let Paul teach you. Read his words and the words of Jesus carefully. Look at the passages in the Tanakh [Old Testament of the Bible] that they reference.

It is our brain and our emotions and our human desire and our spiritual desire that all interact as we near a decision to accept what God has done for us and to live in the light of His plan. One must always count the cost. When we commit our lives to Christ, we are not to become fanatical and overly religious. God is love and truth and grace and gentle, yet strong. And we are to exhibit these traits to the extent that we can. We will NOT become sinless in this life. Do not make that mistake. We ask God to help us diminish our sin and overcome it. After all, we will suffer the consequences of poor actions, ideas, and decisions throughout our life. But it is not our primary focus. Too many Christians behave like god's little sin-finders. It is human (sinful) nature to want to feel better about ourselves at the expense of others. Unfortunately, many groups of confessing Christians apply pressure to join them in their attack of certain people. Our commitment is to love God and to love others. Live with integrity. Act with integrity. Represent God with integrity. Our greatest mission once we realize that heaven is real and his plan is precise, is to help as many as possible to find the same reality in Jesus. It can be scary to think of professing Jesus as the one in whom we place our trust. Unfortunately, one of the biggest reasons this can be intimidating is that the name of Jesus today is often associated with attitudes and beliefs he would not condone. It is the reason why we need to act with as much integrity as possible. We should never get too hung up on the letter of the law. It is the spirit - the sincerity of our heart - that matters. So do not worry about formality. God has always looked on the heart and been focused on what we truly believe in our core.

If you are someone ready to make the commitment, this might be a model for a prayer to God:

Dear God:

I recognize my human condition and that I act well, and that I act poorly. I believe in your plan that can allow me to spend eternity with you and to be led and cared for by you in this current life. So it is by faith that I confess I fall well short of perfection. My hope is not in my ability to live appropriately at all times. My hope is in your fulfillment of the Law and of the plan on my behalf. I ask your forgiveness for my infractions, and I accept the provision made through the sacrifice of Jesus. I dedicate my life to you. Please give me your Holy spirit to guide, nudge and comfort me through the rest of my days here. Please help me grow in wisdom and help me make a difference in the lives of others. Protect me and help me be strong in a world that is both beautiful and dangerous. Help me not to be overly burdened with guilt or other emotions that can stunt my growth and joy in this life. Help me to find a healthy balance between work, service and leisure. Help me to find healthy sources with which to grow. Please help me to give my time and money to important causes and to take responsibility for people around the globe who are suffering. I am willing to work hard. Please provide the opportunities and means to meet my financial needs and obligations.

I know that life can get hard. Help me to struggle well in those times. You know me better than I know myself. Please direct my path so I can use the gifts you have given me and so I can receive the godly desires of my heart. Thank you for all you are doing for me. I ask all of this in the name of the Son, Jesus Christ.

Here are some reminders to ourselves:
We should not expect perfection. We should not need life to be exciting or consistently emotional. We should not get caught up in conspiracy theories that make us look like fools to the rest of the world. We have a message for the world and should not give others reasons to avoid Christianity. Paul compares the Christian life to a race – a very long race. Some days it is easy. Some days it is hard. Sometimes we reflect and realize we are experiencing great faithfulness and effectiveness. Other times we realize we have not been very productive. Let's grow in discipline. Try not to let fear or guilt take hold of you. Seek first God's kingdom, and so much will be provided for you. But also know that some of us will suffer greatly. Do not get bitter. Always put this short life into the perspective of the life that is to come. And it will come. God is real as are his promises. While it is appropriate to be concerned with our personal circumstances, we should routinely consider the plight of others and how we might be of assistance and encouragement. The greatest fulfillment in life tends to come from times in which we devote ourselves to others with little concern of our gain or reward. Every person matters and should be respected. If we cannot truly learn something valuable from another person, we may not be looking hard enough or our ignorance and EBS may be getting in the way.

We should enjoy the good times and treat ourselves and those we love well. We should share and serve regularly and from a heart of love and gratitude; and we should find time to have fun, and experience this planet, time with loved ones, and the things that bring us joy. Struggle well through all of life's challenges.

May we keep our hearts and spiritual eyes on Christ and on an eternity that will absolutely amaze us. God's ability to dazzle and provide with many dimensions of space and time will make any desirable activity in this life seem trivial. We should respect others, including people we may not understand. God truly loves every people group on the planet. We should find ways to share the plan of God with others, while recognizing the need to earn respect first. Let God mold us into the persons we truly want to be. He gives us the desires of our hearts. Sometimes the road to those desires takes years and requires some difficult experiences. Be ready for anything. And love and respect the journey.

Finally, this book can help with a perspective from which to grow. Of all the resources to rely upon, it is the teachings of Jesus and Paul in the New Testament that are foundational. Read with literary integrity. The wisdom is there. It may not follow conventional thinking today. The teachings of Jesus certainly did not always follow the convention of his day. Read carefully and stay humble. We will meet and get to know billions of beautifully transformed persons in the next life. Until then, enjoy the good, the bad, the mundane, and the awe of the journey. Struggle well.

References

Chapter 1:

1 Daniel Kahneman. Thinking, Fast and Slow. Farrar, Straus, and Giroux, 2013. Pages 39 – 49.

2 Sophia Smith Galer. "The Accidental Invention of the Illuminati Conspiracy" BBC.com BBC Future August 9, 2017
https://www.bbc.com/future/article/20170809-the-accidental-invention-of-the-illuminati-conspiracy

Chapter 2:

1 Dr. Fazale Rana. "Do Self-Replicating Protocells Undermine the Evolutionary Process?" Reasons to
Believe. November 12, 2015 http://www.reasons.org/articles/Do-Self-Replicating-Protocells-Undermine-the-Evolutionary-Theory>

2 Ruth Marlaire. "NASA Ames Reproduces the Building Blocks of Life in Laboratory" NASA. Solar
System and Beyond, March 3, 2015
       www.nasa.gov/content/nasa-ames-reproduces-the-building-blocks-of-life-in-laboratory

3 Pearce, Ben K.D. and Pudritz, Ralph E. "Meteorites and the RNA World: A Thermodynamic Model of
Nucleobase Synthesis within Planetesimals." Astrobiology 16 (Nov 16): 853 - 872

4 Ibid

5 Peter D. Ward and Donald Brownlee. Rare Earth: Why Complex Life Is Uncommon in the Universe.

Copernicus Books, 2004. Page 98.

6   Viviane Richter. "The Big Five Mass Extinctions" Cosmos Paleontology
      cosmosmagazine.com/paleontology/big-five-extinctions

7   Hugh Ross. Improbable Planet: How Earth Became Humanity's Home. Baker Books, 2016. Page 169.

8   Ibid, 167, 168

9   Hugh Ross. Why the Universe Is the Way It Is. Baker Books, 2008. Pages 44 – 47.

10   Peter D. Ward and Donald Brownlee. Rare Earth: Why Complex Life Is Uncommon in the Universe.
      Copernicus Books, 2004. Page 42.

11   Hugh Ross. Improbable Planet: How Earth Became Humanity's Home. Baker Books, 2016. Page 38,39

12   Ibid

13   Ibid, 112

14   Kurt O. Konhauser et al. "The Archean Nickel Famine Revisited," Astrobiology 15 (October 15): 804

15   Hugh Ross. Improbable Planet: How Earth Became Humanity's Home. Baker Books, 2016. Page 123.

16   The Drake Equation. "What do we need to know about to discover life in space?" SETI Institute

References

http://www.seti.org/drakeequation

17 Mike Wall, Space.com Senior Writer, "NASA Finds 1284 Alien Planets, Biggest Haul Yet, with Kepler Space Telescope" Space.com Science & Astronomy, May 10, 2016
http://www.space.com/32850-nasa-kepler-telescope-finds-1284-alien-planets.html

18 Ibid

19 Nola Taylor Redd. "New Moon-Formation Theory Also Raises Questions About Early Earth" Smithsonian.com. Science-Nature, August 12, 2016
http://www.smithsonianmag.com/science-nature/new-moon-formation-theory-also-raises-questions-about-early-earth-180960077/

20 Hugh Ross. Why the Universe Is the Way It Is. Baker Books, 2008. Pages 80, 81.

21 Ibid

22 Jesse Emspak. "The Moon Was Formed in a Smashup Between Earth and a Near Twin: But solving one puzzle of lunar origins has raised another linked to the abundances of tungsten in the primordial bodies" Smithsonian.com, April 8, 2015
http://www.smithsonianmag.com/science-nature/moon-was-formed-smashup-between-earth-and-near-twin-180954915/

23 Elizabeth Howell. "Reference: Shoemaker-Levy 9: Comets Impact Left Its Mark on Jupiter," Space.com Science & Astronomy, February 19, 2013.
http://www.space.com/19855-shoemaker-levy-9.html

Chapter 3:

1 Zondervan NIV Study Bible, Genesis 11:10 study notes. Grand Rapids, MI: Zondervan Corporation, 1985
2 Strong's Exhaustive Concordance of the Bible, Updated and Expanded Edition. Word 3117 [yowm].
Peobody, Massachusetts: Hendrickson Publishers, Inc, 2007

Chapter 4:
1 Herodotus, Hist. i.128; iii. 132.2, 159.1

Chapter 5:
1 Matt Williams. "A Universe of 10 Dimensions" Universe Today, November 7, 2016
https://www.universetoday.com/48619/a-universe-of-10-dimensions/

2 Hugh Ross. The Genesis Question: Scientific Advances and the Accuracy of Genesis. NavPress and Reasons to Believe, 2001. Page 17 - 25.

3 Ibid

4 Ibid, 27 - 34

5 Ibid

6 Ibid

Chapter 6:

1 Herodotus, Hist. i.128; iii. 132.2, 159.1

Chapter 7:
1 Fact sheet. "Children: reducing mortality" World Health Organization, updated September 2016
        http://www.who.int/mediacentre/factsheets/fs178/en/
2 Ibid

Chapter 8:

1 T.C. Mitchell, "Geology and the Flood," in New Bible Dictionary, 2nd ed., eds. J.D. Douglas, et al.
(Wheaton, IL: Tyndale, 1982), pages 382-383

2 Ghosh, Pallab. "'First Human' discovered in Ethiopia". BBC News website, Science & Environment.
        4 March 2015 < http://www.bbc.com/news/science-environment-31718336>
3 Krulwich, Robert. "How Human Beings Almost Vanished From Earth In 70,000 B.C."
Krulwich wonders, npr CPR NEWS. October 22, 2012
http://www.npr.org/sections/krulwich/2012/10/22/163397584/how-human-beings-almost-vanished-from-earth-in-70-000-b-c.

4 Krulwich, Robert. "How Human Beings Almost Vanished From Earth In 70,000 B.C. Krulwich wonders,
npr CPR NEWS. October 22, 2012
http://www.npr.org/sections/krulwich/2012/10/22/163397584/how-human-beings-almost-vanished-from-earth-in-70-000-b-c.

5 Zondervan NIV Study Bible, Genesis 11:3 study notes. Grand

Rapids, MI: Zondervan Corporation,
1985
6 Zondervan NIV Study Bible, Genesis 11:4 study notes. Grand
Rapids, MI: Zondervan Corporation,
1985
7 Ker Than. "On Way to New World, First Americans Made a 10,000-
Year Pit Stop". National
Geographic news. February 27, 2014
http://news.nationalgeographic.com/news/2014/04/140227-native-americans-beringia-bering-strait-pit-stop/

8   Hugh Ross. The Genesis Question: Scientific Advances and the
Accuracy of Genesis. NavPress and Reasons to Believe, 2001. Page
123 - 125.

9   Ibid, 119 – 123

INDEX

Abraham, 34, 44, 50, 55, 61, 67, 91, 109

Adam, 27, 105

afterlife, 40, 118

atheist, 69

bacteria, 9

baptism, 115

Bering Land Bridge, 87

blasphemed, 29

building blocks of life, 8

Caesar Augustus, 94

Cain and Abel, 88

cities of refuge, 98

cloud cover, 25

Covenant, 45, 61, 67

created, 58

crucifixion, 20, 111

curse, 35, 66, 111

cytosine, 9

David, 46, 49, 91

death, 27, 116

dimensions

   Time, 27, 53, 55, 104

    Space, 26, 52, 55, 103

DNA, 8

Drake, Francis, 12

donkey, 62, 94

drunkards, 70

Earth, 6 - 11, 23 - 26, 57, 98, 103, 117

EBS, 20, 101

eclipse, 14

ecosystems, 10, 99
Eden, 88
electromagnetic field, 11
elements, 11
empathy 83
exoplanets, 13
extinctions, 12
flood, 25, 84, 107
Gandhi, 35
generations skipped, 23, 108
Gentiles, 29
giant planets, 15
gluttons, 70
Golden Age of Islam, 38
gravitational force, 14
group think, 7, 21
guilt, 81
habitable zone
  Galaxy, 11, 13
  Solar system, 12, 13
heart, 41, 116
heaven(s),41, 57, 58, 116
hell, 39, 79
heterosexuality, 29
hominins, 85
homosexuality, 29
horns of the altar, 92
House of Wisdom, 21
humans, 84, 86
hung on a tree, 66, 110
ideology, 101
Isaiah, 65
Islam, 19, 37
Jerusalem, 62, 63, 92, 94

Index

Josephus, 46

Judaism, 19

judgment, 29, 32, 41, 65, 117

lamb, 61, 65, 92, 109, 110

Law, 33, 34, 40, 41, 46, 59, 62, 66, 67, 69, 110

LGBTQ, 22, 28, 30, 33

life, 6, 7, 10, 13, 15

luminosity, 10, 12

Melchizedek, 49, 61, 91, 109

Mesopotamian Region, 25, 85, 89

Messiah, 48, 57, 63 – 65, 109, 118

military, 73

moon, 13, 14

Moses, 45, 49, 60, 61, 70, 92, 110, 112, 115

Muslims, 22, 37

My God, my God, 20, 47, 93

Nod, 88

nucleotide bases, 9

pain, 1, 38, 72, 77

palm branches, 63, 94

Passover, 62, 92

Paul, 47, 52, 59, 66, 69, 72, 111, 114, 116

prebiotics, 9

probability, 7

protocells, 8

radical Christianity, 37

radical Islam, 37

radioactive
 elements, 11

Ross, Hugh, 15, 58

sacrifice, 61, 92, 109

Saul of Tarsus, 17, 47, 66

Scripture, 18, 91, 101

seed, 69

Index

self-replicating material, 8
self-righteousness, 33, 35
stretches out the heavens, 57
supernovae, 88
supporting life, 13
Tanakh, 19, 20, 59, 86, 99, 111
telomerase, 88
they, 30-32
volcano, 86
water levels, 87

Index

NOTES:

Index

NOTES: